Modeling and Simulation of Logistics Flows 3

Series Editor
Jean-Paul Bourrières

Modeling and Simulation of Logistics Flows 3

Discrete and Continuous Flows in 2D/3D

Jean-Michel Réveillac

iSTE

WILEY

First published 2017 in Great Britain and the United States by ISTE Ltd and John Wiley & Sons, Inc.

ISTE Ltd
27-37 St George's Road
London SW19 4EU
UK

www.iste.co.uk

John Wiley & Sons, Inc.
111 River Street
Hoboken, NJ 07030
USA

www.wiley.com

Library of Congress Control Number: 2016956897

British Library Cataloguing-in-Publication Data
A CIP record for this book is available from the British Library
ISBN 978-1-78630-108-6

Contents

About This Book ix

Introduction xv

Chapter 1. Computer Simulation of Discrete Flows 1

1.1. Introduction 1
1.2. Worked example 1
1.2.1. Map of the resort 2
1.2.2. Problem statement and design brief 3
1.3. Setting up the project in the ExtendSim 9 software 5
1.3.1. Definition of the principal parameters 5
1.3.2. Designing the model and inputting constraints 7
1.3.3. Definition of flows 22
1.3.4. Running the simulation 22
1.3.5. Creation and allocation of resources 24
1.3.6. Rerunning the simulation 28
1.3.7. Generating a report and analysis 29
1.3.8. Development, enhancement and improvement 31
1.3.9. Hierarchy 38
1.3.10. Appearance design 40
1.4. Conclusion 44

Chapter 2. Simulation of Mixed Flows 47

2.1. Mixed Flows 47
2.2. An example of modeling mixed flows 48
2.2.1. Problem statement and specifications 48
2.3. Creating and inputting the project in ExtendSim 52
2.3.1. Definition of the principal parameters 52

2.3.2. Soda production and bottling 53
2.3.3. Transport, carbonation and labeling 80
2.3.4. Packaging and storage 85
2.3.5. Maintenance and cleaning. 93
2.3.6. Finishing touches 98
2.4. Conclusion 108

Chapter 3. 3D Flows and Evacuation Simulation 109

3.1. 3D flows 109
3.2. The Pathfinder software 110
3.3. Evacuation of a building with PathFinder 111
3.3.1. Importing and formatting the first floor plans 113
3.3.2. Creating the different first floor rooms 117
3.3.3. Creating the first floor doors 120
3.3.4. Populating with occupants 122
3.3.5. Simulation and results for the first floor evacuation 123
3.3.6. Incorporating furniture. 126
3.3.7. Importing and formatting the second floor plans 128
3.3.8. Creating rooms, doors and populating with occupants 129
3.3.9. Creating the stairs. 130
3.3.10. Simulation and results for evacuation of the whole building. . . . 134
3.4. Extensions 146
3.4.1. Moving to SFPE mode. 146
3.4.2. Groups of occupants 148
3.4.3. Managing the elevators 148
3.4.4. Creating viewpoints 154
3.4.5. Creating camera tours 156
3.4.6. Further possibilities. 158

Chapter 4. 3D Flows, Distribution and Warehousing. 159

4.1. Product distribution 159
4.2. The FlexSim software. 159
4.3. Basic concepts of the FlexSim software 160
4.3.1. General appearance of FlexSim 160
4.3.2. Libraries 162
4.3.3. Mouse-based functions 164
4.3.4. Connections between objects 165
4.4. Worked example. 166
4.4.1. Description of the warehouse. 167
4.4.2. Warehouse operation. 168
4.4.3. Modeling stage 1 170
4.4.4. Modeling stage 2 178

4.4.5. Modeling stage 3 184
4.5. Detailed flow and task executer management. 194
4.5.1. Generation of containers with several types of content. 194
4.5.2. A fixed resource for task executers 198
4.5.3. Shared task executers 200
4.5.4. Pulled and pushed flows and more. 204
4.5.5. Naming items 210
4.5.6. Timetables, groups and resources 216
4.6. Experimenter 231
4.6.1. Constructing the model 231
4.6.2. Adding the dashboard 232
4.6.3. Configuring the Experimenter 235
4.7. Concluding remarks. 239

Conclusion. 241

Glossary 245

Bibliography. 251

Index 259

About This Book

There are already several works about logistics, operational research, decision support, the theory of graphs, dynamic programming, etc., but few of them gather all of these domains together by proposing an overall vision that focuses less on pure and hard mathematical aspects, without totally ignoring them, while offering numerous practical exercises.

This book is one of three volumes. This first volume tackles theoretical aspects with corrected exercises for each chapter, finishing with a presentation of the principal software systems dedicated to operational research (OR) and logistical simulation. The second and third volumes are dedicated to practice and specialized applications of software programs.

Most of the studies proposed here are able to be completed using a simple calculator, a sheet of paper and a pen or even with the help of a spreadsheet on Microsoft Excel, Apache OpenCalc, Apple Numbers, etc.

The presented techniques and their uses are multiple, yet I am sure that a student, a software programmer, a developer, a technician, an engineer, an IT specialist, a decision-maker and you, the reader, will find practical applications that were unexpected in your professional or even personal life.

Intended public

This work is designed for all those who encounter logistical problems linked to flux management, decision support, optimization of journeys or rounds, research for an aim when confronted with multiple constraints, creation of dashboards, relevant simulations, etc.

The works presented require a minimum level of mathematical knowledge, and a post A level student in science or economics will not encounter major difficulties. I tried to maintain simplicity and go straight toward the objective in the theoretical approach without going into great demonstrations, which to me do not seem necessary.

In terms of the practical exercises, on a laptop, tackled in Volumes 2 and 3, a good knowledge of the exploitation system (track, records and lists, files, names, extensions, sheet, movement, etc.) will prove essential.

Since a few works use a spreadsheet, it is thus necessary to master the basic functionalities of this type of software program. It will also be convenient to know the primary use of pivot table data manipulation tools.

If we know about *visual basic application* (VBA) language or its equivalent, we can easily understand, improve, enrich and create new solutions to certain problems.

Lastly, if we understand the basic systems for managing data and relational algebra, then we will be at ease in every domain explored.

Organization and contents of the book

This work is composed of three volumes:

– Volume 1: Theory and fundamentals;

– Volume 2: Dashboards, traffic planning and management;

– Volume 3: Discrete and continuous flows in 2D/3D.

Volume 1 presents an introduction followed by 10 chapters and a conclusion:

– approach for logistics;

– an overall view of operational research;

– basics of the theory of graphs;

– calculation of optimal routes;

– dynamic programming;

– planning and scheduling with PERT and MPM;

– calculation of waves in a network;

– hiding trees and tours;

– linear programming;

– modelization of route traffic;

– diverse software programs for RO and simulation of logistical flux.

Here, we will find the fundamental concepts needed in order to understand the second volume. Numerous examples accompany the theory and each chapter ends with a series of exercises with their solutions.

The conclusion, as indicated by the name, tries to establish a picture of the current state of theoretical logistics and its future development.

Appendices 1 and 2 bring a few other elements. We will find them in the following order:

– table of the law of the normal reduced center;

– a presentation and a mini-manual dedicated to the calculation software program GeoGebra.

Volume 2 starts with an introduction completed with four chapters that put into practice the software tools in cases of practical application in order to finish with a conclusion:

– the different tools used in this volume;

– operational research with a spreadsheet;

– dashboards with a spreadsheet;

– scheduling and planning with a project manager;

– simulation of route traffic.

The conclusion presents new functionalities that should appear on spreadsheets and project management systems as well as the evolutions and points of similarity between traffic simulators and new infrastructures that emerge in traffic networks.

Appendix 1 is dedicated to the installation of a solving tool in Microsoft Excel. Appendix 2 is consecrated to the installation of Java development kit.

Volume 3 starts with an introduction followed by four chapters dedicated to the modelization and simulation of flux in a 2D or 3D environment. Each case is different and taken from situations encountered in reality. A conclusion concludes this Volume 3:

– different software programs used in this third volume;

–simulation of discreet computerized flows;

–simulation of mixed flows;

–3D flows and evacuation simulation;

–3D flows for transporting and storing.

The conclusion conveys the future evolutions of software programs and their integration into society. At the end of each volume, we will find a bibliography and a list of Internet links. A glossary is also available that will elaborate on certain acronyms and some very specific terminology surrounding logistics and operational research.

Conventions

This book uses the following typographic conventions:

– *italics*: used exclusively for important terms used for the first time in the text, one can generally be found in the glossary at the end of the work mathematical terms, comments, equations, expressions or variables present in the theoretical and practical chapters among examples and exercises;

– *(italics)*: these are the terms in English or in foreign languages;

– UPPER CASE: these are reserved for names of windows, icons, buttons, files or lists, menus or submenus. This can also be elements, options or commands present in the window of a program;

– `courier`: this font is used for VBA code lines. These lines can end with the symbol ↵, which implies a return to the obligatory line when inputting.

Any comments are signaled by the presence of a keyword: COMMENT. They will complete the explications already provided. Theorems are signaled by the keyword: THEOREM. The figures and tables have captions to further their understanding.

Vocabulary and definition

Like for all techniques, logistical optimization tools have their own vocabulary. Words, acronyms, abbreviations and specific names that are not always familiar; this is the role of the Glossary found at the end of the book.

Acknowledgments

It is my particular duty to thank the team at ISTE, and my editor Chantal Ménascé who had confidence in me, Jacqueline Gélinier from the company 1point2, distributor of ExtendSim and PathFinder software programs, Clair Augsburger of FlexSim France, not forgetting my dear friend, Pascal Mauny, director of IUT of Chalon-en-Saône and lecturer at the University of Bourgogne, for the time and attention granted to me for writing the preface.

Finally, I would like to thank my wife, Vanna, who has supported me throughout writing this book.

Jean-Michel RÉVEILLAC
November 2016

Introduction

Logistics and simulation of different flows

The last chapter of Volume 2 of this book concluded with a presentation of different road traffic-based examples, which were modeled and simulated using several pieces of software.

In this third volume, we will encounter simulation problems dealing with a variety of different flows (for example of skiers, pedestrians, crates, boxes, pallets, liquids, bottles, etc.) in discrete, continuous or mixed formats.

In relation to what was presented previously, new ideas and concepts will be developed while making use of some of the theories present in Volume 1.

We will continue to deal mostly with problems arising from real situations encountered in the field, in some cases with a simplified design to facilitate their modeling and make them more accessible to the user. The goal is to provide the basics and tools that will allow, in time, for more complex problems to be tackled.

We will work with the flow simulators ExtendSim, PathFinder and FlexSim, in turn.

In the first chapter, we will start by creating a model simulating the ski slopes of a French winter sports resort with its ski lifts, pistes and skiers. This example will demonstrate the feasibility of creating a model associated with discrete flows in a non-industrial environment, where simulation can, for example, contribute to characterizing the installation of a future ski lift. Given the considerable construction costs of such an installation, mistakes in decision-making must be avoided – a virtual model allows comparison of multiple options with the design brief in order to decide upon the best solution(s).

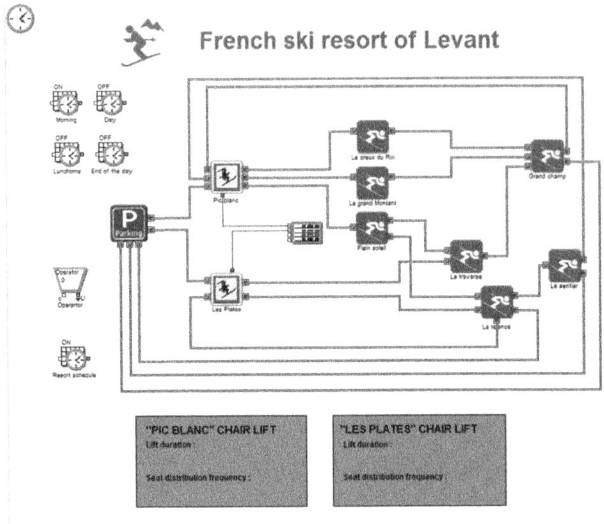

Figure I.1. *The model of the ski station. For a color version of this figure, see www.iste.co.uk/reveillac/modeling3.zip*

Within industry, logistics flows are often mixed, which in the second chapter leads us to deal with an example of a production and packaging line for fizzy drinks using multiple throughput control, mixing, transport, bottling, warehousing and storage sub-systems. The model will combine liquids, which are continuous, with bottles, pallets and crates, which are discrete.

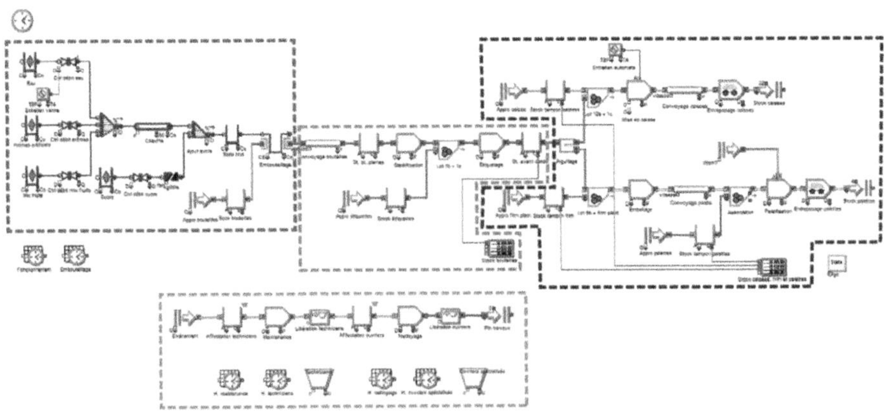

Figure I.2. *Model of the production of fizzy drinks. For a color version of this figure, see www.iste.co.uk/reveillac/modeling3.zip*

In the third chapter, which covers simulation, we approach the problem of simulating flows in 3D by developing a model simulating the evacuation of people from a building. We will see how to import and integrate a building blueprint into a model, and then how to put the different actors (pedestrians) in place, as well as how to manage their behavior.

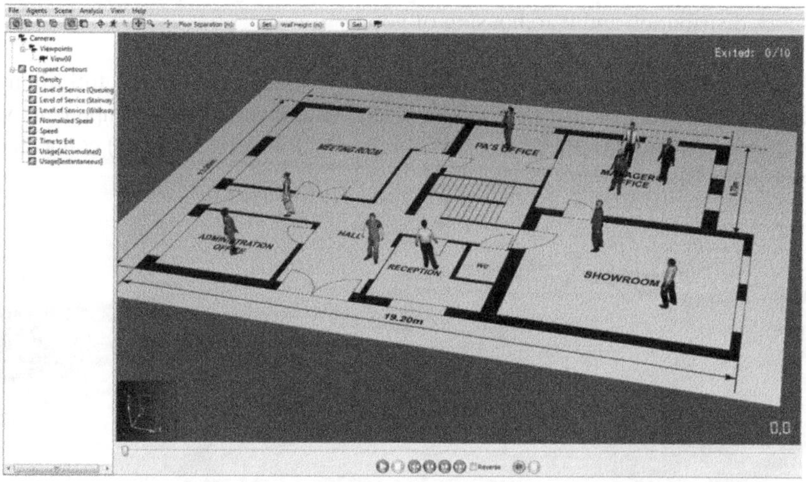

Figure I.3. *The ground-floor plan of our building with its different occupants*

We will incorporate furniture, stairs and a lift to obtain an even more realistic simulation.

Figure I.4. *A running simulation of the center of the building in our example. For a color version of this figure, see www.iste.co.uk/reveillac/modeling3.zip*

Finally, we will generate different graphics from the text-based result tables summarizing all of the parameters of the simulated scenario.

A very active domain in the simulation of logistic flows is the shipping and storage of goods. We will deal with these aspects in the last chapter with an example of a semi-automated platform covering the unloading, unpacking, relabeling, transport, distribution and warehousing of goods.

To carry out these tasks, we will use classic resources such as forklift truck drivers and operators in terms of human resources and automated components with roller conveyors and *ASRS* (Automated Storage and Retrieval Systems) connected to rows of multi-storey storage racks.

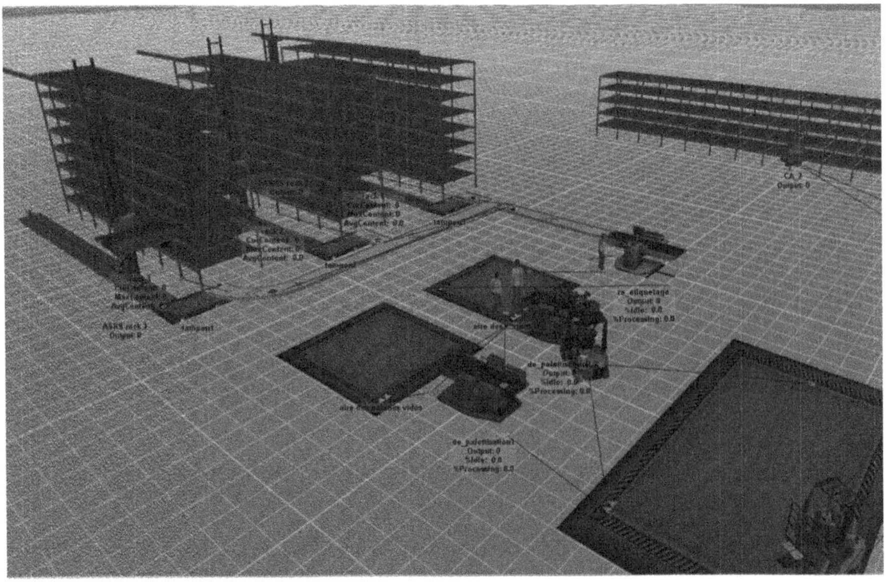

Figure I.5. *The warehouse platform. We can see the forklift trucks and their drivers, the operators in yellow and, at back-left, the storage racks with their ASRS. For a color version of this figure, see www.iste.co.uk/reveillac/modeling3.zip*

The construction and simulation will be conducted in 3D under externally imposed, precise conceptual and organizational constraints, which could be varied to improve the efficiency of the platform.

To measure and visualize the fundamental variables linked to the running of our model, we will design dashboards and reports allowing visualization of the results produced by the different simulations.

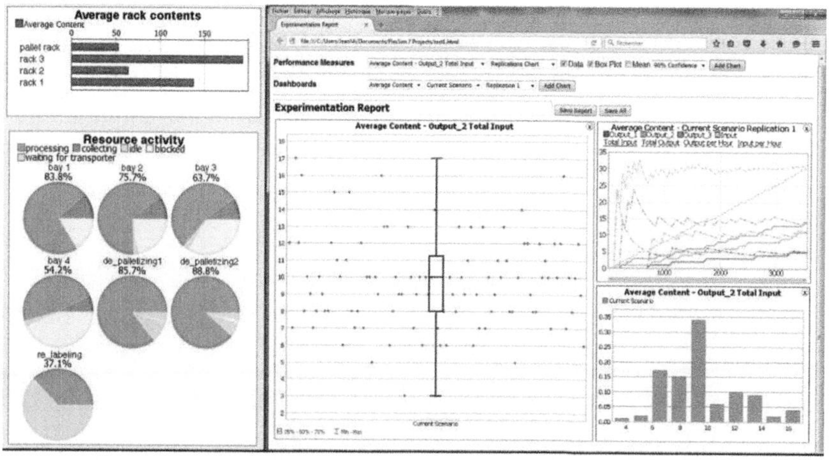

Figure I.6. *Simulation dashboards and report. For a color version of this figure, see www.iste.co.uk/reveillac/modeling3.zip*

To conclude, we will mention some specific situations and see the relevant solutions that we can put in place to meet their requirements.

1

Computer Simulation of Discrete Flows

1.1. Introduction

The simulation of computerized flows gained its reputation with the development of the field of logistics and as a result of the computational power now available on personal computers.

All types of flows can be modeled, whether they are discrete or continuous, and many businesses use simulation tools instead of physical systems or prototypes and the inevitable costs they entail.

This is not to say that the software market has abundant applications, in fact there are relatively few, but the editors have built up considerable experience over the past few years and can suggest uses covering most sectors (see section 10.4 of Volume 1).

Historically, the first pieces of simulation software appeared in the 1990s (Scitor Process, ExtendSim, etc). At that time, the computing power that was available limited the capability of the software, and their usability was far from convincing.

The main body of this chapter deals with a non-industrial example, chosen intentionally to demonstrate that logistics and the simulation of flows can also be applied in more unusual sectors.

1.2. Worked example

In order to give a glimpse of the range of possibilities covered by computerized flow management, in particular with the ExtendSim 9 software, we will create a simulation modeling the small French winter sports resort 'Levant', with its skiers, ski lifts and pistes.

COMMENT 1.1.–

– The ExtendSim 9 software has been chosen because this is what I teach to my students in the 'Logistics and Transport Management' department at the University of Burgundy – Faculty of Science & Technology and Institute of Technology Chalon-sur-Saône.

– A usable version of the software is available on the publisher's Website (see the Internet links at the end of the book).

– We thank 1point2, the official distributor of the ExtendSim software in France, for providing us with a US version.

1.2.1. *Map of the resort*

Figure 1.1 shows a map of the whole ski resort, with its pistes and ski lifts.

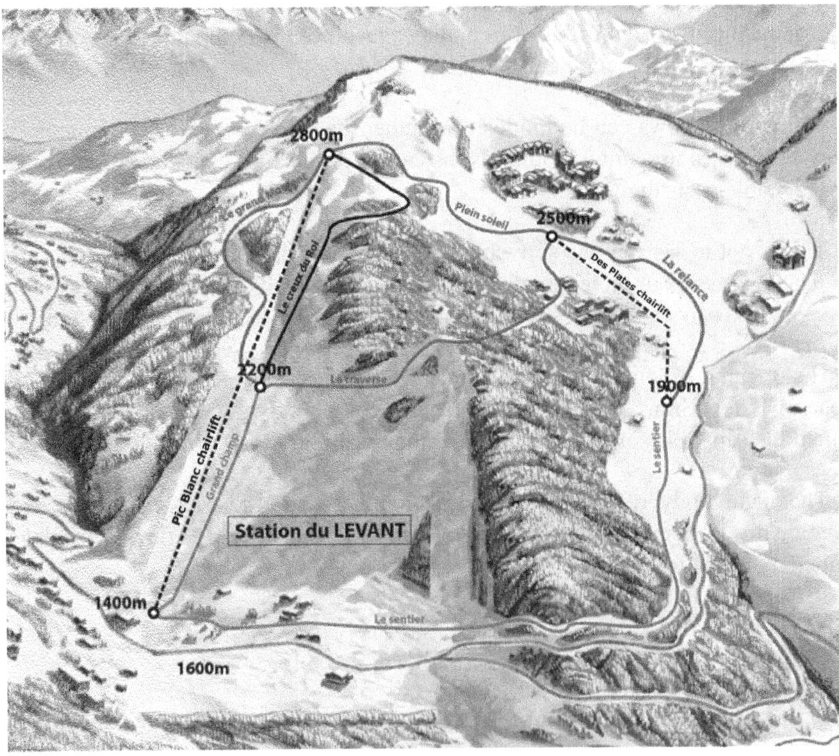

Figure 1.1. *Map of the 'Levant' ski resort. For a color version of this figure, see www.iste.co.uk/reveillac/modeling3.zip*

1.2.2. *Problem statement and design brief*

The full specifications to be incorporated are as follows:

– Resort opening times: 9:00 am–5:30 pm.

– Arrival frequency of skiers at the ski lifts, in seconds:

- 9:00 am–10:30 am: mean = 5 s, standard deviation = 1.5 s;
- 10:30 am–12:30 pm: mean = 14 s, standard deviation = 6 s;
- 12:30 pm–1:45 pm: mean = 9 s, standard deviation = 4s;
- 1:45 pm–2:30 pm: mean = 14 s, standard deviation = 6 s;
- 2:30 pm–5:30 pm: mean = 20 s, standard deviation = 7.5 s.

– Ski lift specifications:

- 'Pic Blanc' chairlift:
  – capacity per chair: 4 skiers;
  – lift duration: 9 min;
  – frequency of chair arrival: 18 s;
  – personnel: 1 operator.
- 'Plates' chairlift:
  – capacity per chair: 3 skiers;
  – lift duration: 6 min;
  – frequency of chair arrival: 15 s;
  – personnel: 1 operator.

– Workforce required:

- two operators (1 per chairlift) who work from 9:00 am to 5:30 pm;
- hourly cost of 1 operator: $18.25/h.

– Skier descent time by piste:

- 'Grand champ': mean = 5 min, standard deviation = 2.5 min;
- 'La traverse': mean = 9 min, standard deviation = 3.5 min;
- 'Le creux du Roi': mean = 6 min, standard deviation = 2.5 min;
- 'Le grand Montant': mean = 7 min, standard deviation = 2 min;
- 'Plein soleil': mean = 6 min, standard deviation = 1.5 min;

- 'Le sentier': mean = 11 min, standard deviation = 3 min;
- 'La relance': mean = 6.5 min, standard deviation = 1.75 min.

– The probabilities of the different choices skiers make at each decision point are as follows:

- upon arrival at the resort:
  – take the 'Pic blanc' chairlift: 64%;
  – take the 'Plates' chairlift: 36%.
- upon arrival at the foot of the 'Grand champ' and 'Le sentier' pistes:
  – head toward the start of the 'Pic blanc' chairlift: 88%;
  – head toward the parking lot (leave the slopes): 12%.
- upon arrival at the foot of the 'La relance' piste:
  – head toward the 'Le sentier' piste: 60%;
  – head toward the 'Plates' chairlift: 28%;
  – head toward the car-park: 12%.
- upon arrival at the top of the 'Pic blanc' chairlift:
  – head toward the 'Le grand Montant' piste: 35%;
  – head toward 'Le creux du Roi' piste: 20%;
  – head toward the 'Grand soleil' piste: 45%.
- upon arrival at the top of the 'Des Plates' chairlift:
  – head toward the 'La traverse' piste: 62%;
  – head toward the 'La relance' piste: 38%.
- at the foot of the 'Plein soleil' piste:
  – head toward the 'La traverse' piste: 56%;
  – head toward the 'La relance' piste: 44%.

– Assorted constraints and observations:

- evidently, the pistes shown on the map are for descent only, the skiers do not climb up the slopes. The altitudes displayed on the map indicate the direction of descent;

- skiers can reach both chairlifts by road and arrive with a vehicle. A parking lot is available in front of each chairlift;

- when a skier arrives at the foot of the 'Grand champ' and 'Le sentier' pistes, they can decide to stop skiing and return to their vehicle, situated in the 'Pic blanc' parking lot. This is also the case at the foot of the 'La relance' piste, where the skier can return to the 'Plates' parking lot;

- skiers can return to the 'Plates' chairlift by taking the 'Pic blanc' chairlift and then descending the 'Plein soleil' and 'La relance' pistes;

- ski passes are given to the skiers in the form of a swipe card which they wear around their necks;

- to access the chairlifts, the skiers must present their ski pass at the bottom of the lift to a gate machine into which they insert their swipe card, each time through. Two machines are available at the 'Pic blanc' chairlift, and one is available at the 'Plates' chairlift. Each verification system is preceded by a queue which must never be greater than 25 skiers. The duration of the ski pass check for the skier is estimated to have a mean value of 5 s and a standard deviation of 3 s.

COMMENT 1.2.–

– All of these specifications derive from measurements made on location over several days. The specifications related to the technology and the workforce were supplied by the department of the 'Levant' local authority in charge of winter sports.

– This simulation does not take into account the purchase and sale of ski passes to skiers. Nevertheless, each skier should have one in their possession to use the pistes and take the ski lifts.

– The simulation should take place over 1 day, from 8:30 am to 6:00 pm.

– Warning: The schedule management uses a 24 h system (0:00:00 to 23:59:59).

1.3. Setting up the project in the ExtendSim 9 software

We will now construct the simulation in the ExtendSim software. For this, we work through a certain number of stages as presented in section 10.4.4 of Volume 1.

1.3.1. *Definition of the principal parameters*

To start, we define a new model: FILE menu, NEW MODEL.

Once the model is created, we define the time parameters of the simulation: RUN menu, SIMULATION SETUP.

Our simulation is discrete because the flow is composed of skiers (who are indivisible and countable).

The time will be managed according to a calendar, so we check USE CALENDAR DATES and choose 01/04/2015 for the START as defined by the design brief, as well as the start time of the simulation, which is 8:30:00.

By default, the unit of time is minutes (GLOBAL TIME UNITS) because it seems to be most-suited to the different future applications of the simulation.

The END TIME is set to 570 min, which is 570/60 = 9.5 h, because our simulation should cover a day from 8:30:00 to 18:00:00 (Figure 1.2).

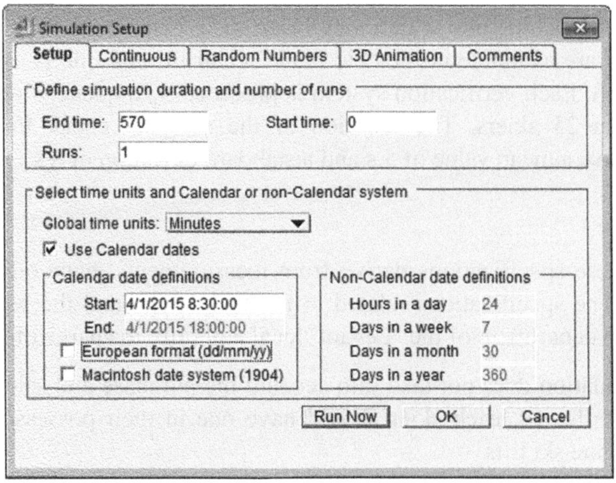

Figure 1.2. *Settings for the simulation*

To include our parameters, we will place the EXECUTIVE block from the 'Item.lix' library in our model, in the top left-hand corner.

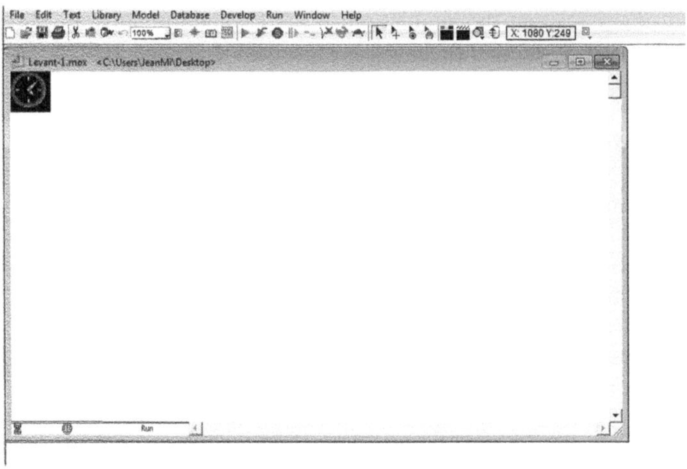

Figure 1.3. *The new model with its EXECUTIVE block*

1.3.2. *Designing the model and inputting constraints*

With the help of different blocks taken from the libraries (files with the extension '.lix'), we will now design and construct the functional organization chart of our simulation. At the same time, we will also input the constraints.

1.3.2.1. *Schedules and generators*

To simulate the arrival of skiers at the resort, we will use the CREATE block ('Item.lix' library) whose function is to create items in order to form a flow which, in our case, will be composed of skiers.

A generator can be associated with a statistical distribution and a time period. In our design brief, we have four different skier arrival frequencies, following a normal distribution (mean and standard deviation), dependent upon four schedules.

We must thus place four CREATE blocks associated with four SHIFT blocks ('Item.lix' library) in our model.

Take, for example the shift from 9:00 am to 10:30 am – a skier arrives, on average, every 5 s, with a standard deviation of 1.5 s, so we parameterize this block accordingly.

We start by creating the schedule which we will call 'morning schedule' (Figure 1.4).

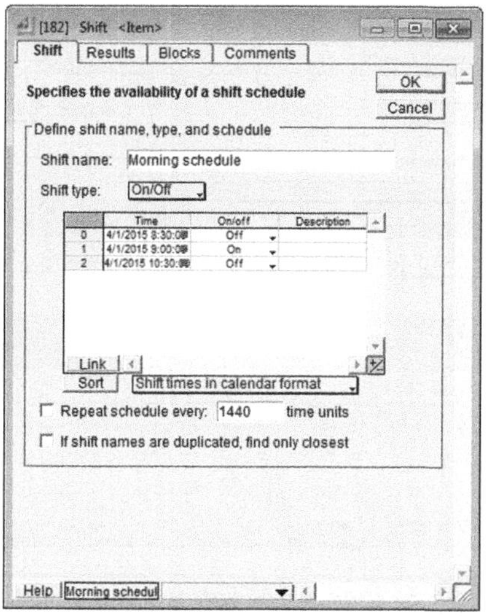

Figure 1.4. *The configuration window for the 9:00 am to 10:30 am schedule*

In Figure 1.4, we can see the name, the timeslot with a start time of 8:30 – OFF, corresponding to the start time of the simulation that we previously input in the SIMULATION SETUP window (section 1.3.1), an arrival time for the skiers, 9:00 – ON and a time at which the frequency changes, 10:30 – OFF.

The schedule is in the calendar format, which is easier to manage in our case because it is what has been chosen in SIMULATION SETUP

COMMENT 1.3.–

– A calendar must always have a start time equal to the start time input in the simulation parameters.

– To add or delete a row, you simply need to click on the green +/– box.

– The ON position indicates the start of an active schedule period, the OFF position indicates the end of one.

– The schedule can be in the numerical format, in which case, according to the unit chosen in SIMULATION SETUP, we specify the number of time units passed since 0 (the time at which the simulation starts).

We can now finalize our CREATE block with the arrival frequency of skiers that we input in the CREATION tab (Figure 1.5).

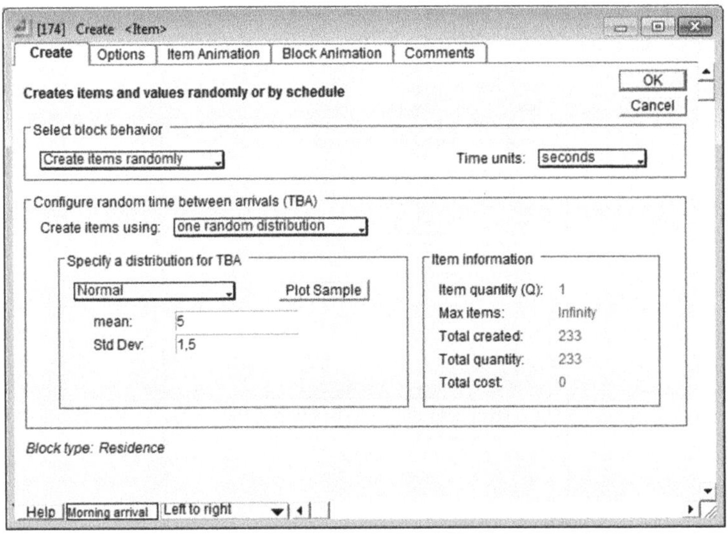

Figure 1.5. *The CREATION tab of the CREATE block for the morning arrivals (9:00–10:30)*

We continue onto the OPTIONS tab, in which we choose the 'Morning shift' calendar that we have just created (Figure 1.6).

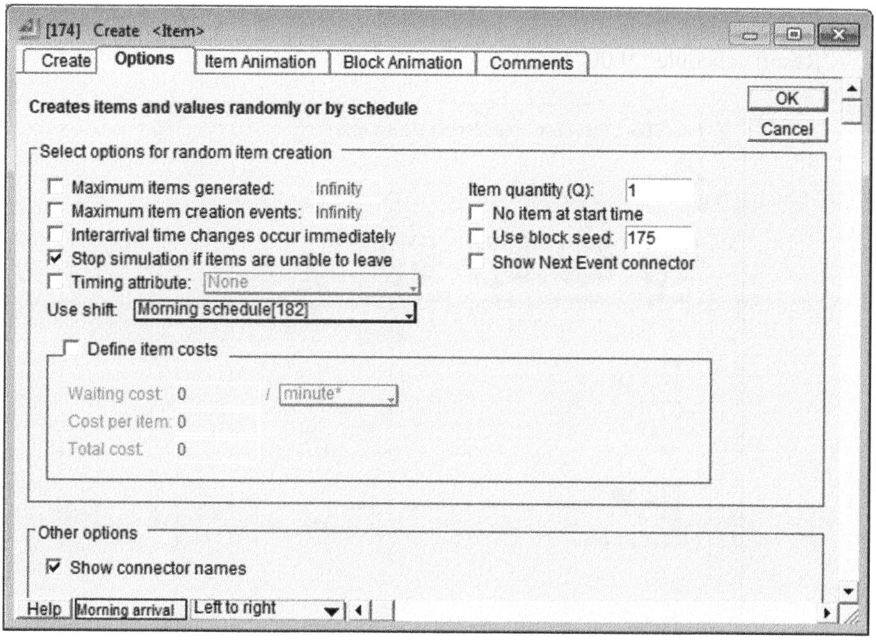

Figure 1.6. *The OPTIONS tab of the CREATE block for arrivals between 9:00 and 10:30. We can see the choice of 'Morning shift' for the option USE SHIFT*

In the same way, we generate three other CREATE blocks, associated with the three other SHIFT blocks.

COMMENT 1.4.– We can group the schedules 10:30–12:30 and 13:45–14:30 in the same SHIFT block since they have the same mean and standard deviation parameters.

We also add another SHIFT block which determines the operating times for the ski lifts and the opening/closing of the pistes, which is 9:00 am–5:30 pm.

In summary, the schedules and generators that we have made are:

– 'Morning schedule' with the 'Morning arrival' generator: 9:00–10:30;

– 'Lunchtime schedule' with the 'Lunchtime arrival' generator: 12:30– 13:45;

– 'Day schedule' with the 'Day arrival' generator: 10:30–12:30 and 13:45–14:30;

– 'End of day shift' with the 'End of day arrival' generator: 14:30–17:30;

– 'Resort schedule': 9:00–17:30.

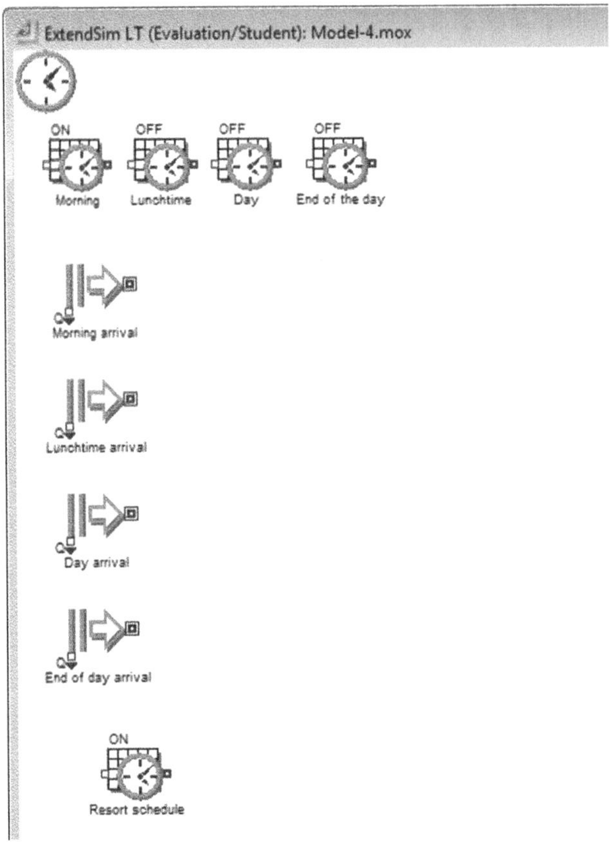

Figure 1.7. *The five SHIFT blocks and the four CREATE blocks*

1.3.2.2. *Ski pass checks*

Our skiers will now be directed toward one of the two ski lifts, the 'Pic blanc' chairlift or the 'Des Plates' chairlift.

Upon arrival at the chairlifts, they will need to present their ski pass to the card reader and go through the turnstile.

Figure 1.8. *Two card readers with turnstiles*

The 'Pic blanc' chairlift has two card readers and the 'Plates' chairlift has one, which we will need to place in front of the queues (QUEUE block, 'Item.lix' library) to receive the skiers.

We are told in the design brief that these queues must not exceed 25 skiers. We cannot formalize this constraint, which is fully dependent upon our simulation, and we should instead treat it as a result of the simulation. We will discuss this further in the report and analysis stage (see section 1.3.7).

In order to make the skiers gather toward our chairlifts, we put in place a SELECT ITEM IN (4 inputs/1 output, 'Item.lix' library), then a SELECT ITEM OUT (which we will name 'Chairlift choice', 'Item.lix' library) so as to take account of the probabilities of each direction, being 64% toward 'Pic blanc' and 36% toward 'Des Plates'.

Another SELECT ITEM OUT (which we will name 'Queue choice') will be placed in front of the queues (QUEUE block) at 'Pic blanc' in order to steer the skiers toward the least congested queue (logically, the skier would move toward the shortest queue).

Since the 'Des Plates' chairlift only has one queue, it should be sufficient to link the QUEUE + ACTIVITY blocks.

The ski pass checks are defined using ACTIVITY blocks. These blocks will be configured following a normal distribution with a mean value of 5 s and a standard deviation of 3 s for each skier (as indicated in the design brief). We will also use the 'Resort schedule', as defined previously.

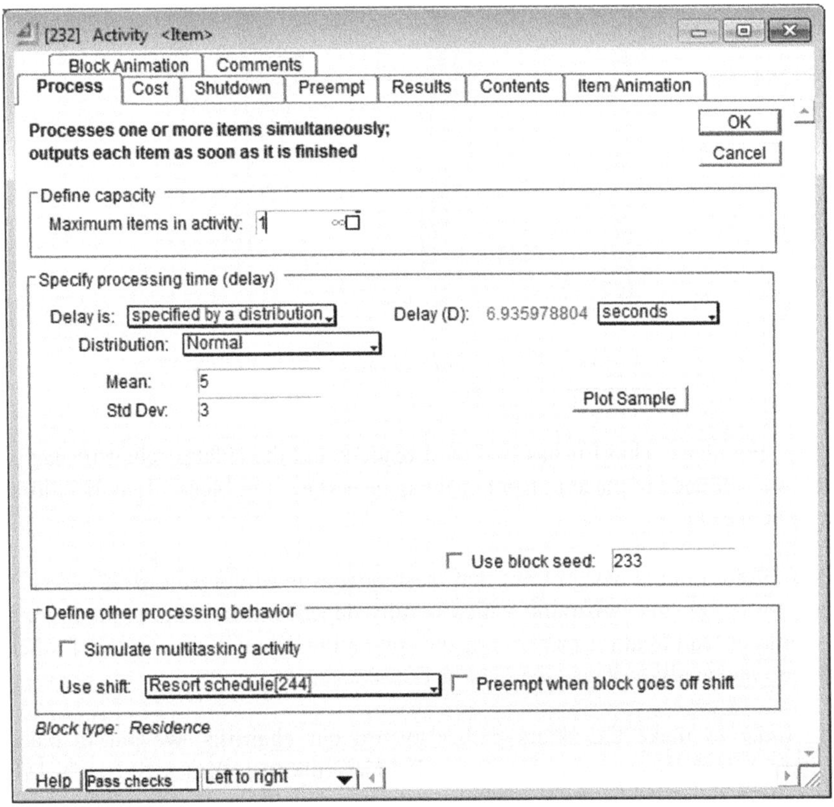

Figure 1.9. *Parameters for the 'Pass checks' ACTIVITY blocks (Distribution: Normal, mean = 5, standard deviation = 3)*

Once all of these blocks are in place, we can connect them together to obtain the model shown in Figure 1.10.

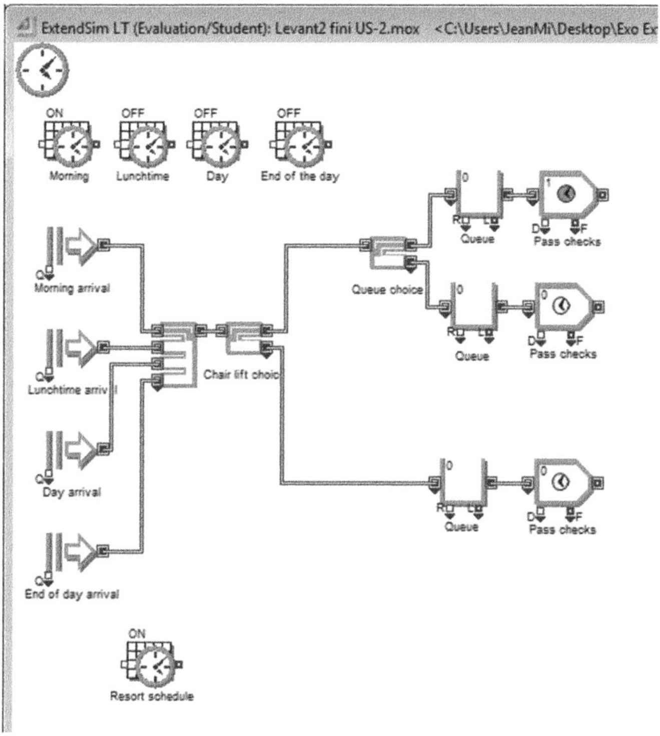

Figure 1.10. *Our model with queues, ski pass checks, the selectors and connections*

COMMENT 1.5.– It is worth regularly saving your model (FILE menu – SAVE) to avoid losing your work, or to be able to return to an earlier version.

We now move on to inputting the different constraints.

On the 'Chairlift choice' SELECT ITEM OUT block, which directs the skiers toward one or other of the chairlifts, we input 0.64 (64%) toward 'Pic blanc' and 0.36 (36%) toward 'Plates'.

For the 'queue choice' selector, we encounter the problem of how to direct the skier toward the least congested queue.

To resolve this problem, we will place a MAX & MIN block, taken from the 'Value.lix' library, in our model. Then, we will direct the value (the number of skiers) contained in of each of the two queues toward it, so that it sends the skier to the most appropriate queue (whichever is shortest).

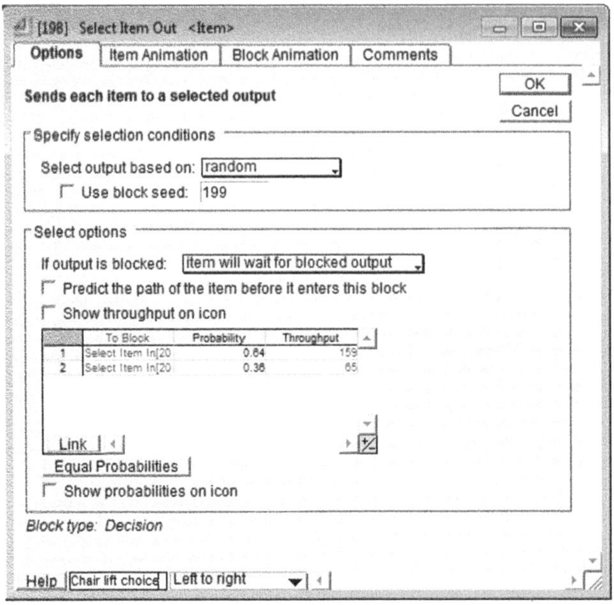

Figure 1.11. *The parameters of the 'Chairlift choice' selector (Probability: 0.64 and 0.36)*

The configuration of the MAX & MIN block is given in 1.12: OUTPUT THE: MINIMUM VALUE.

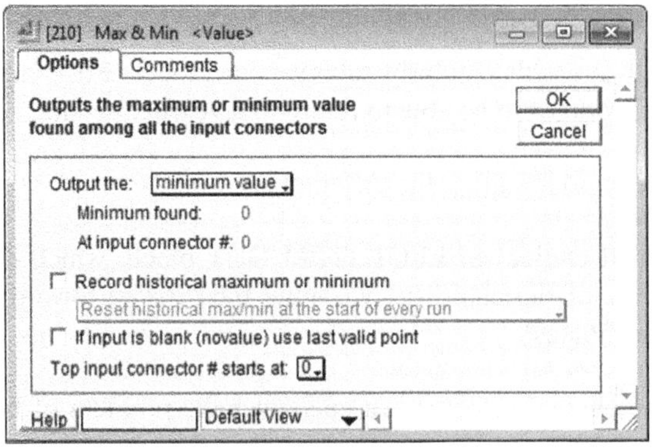

Figure 1.12. *The configured MAX & MIN block (set to MINIMUM VALUE)*

The configuration of the 'Queue choice' selector is shown in Figure 1.13: CHOOSE EXIT: BY SELECT CONNECTOR.

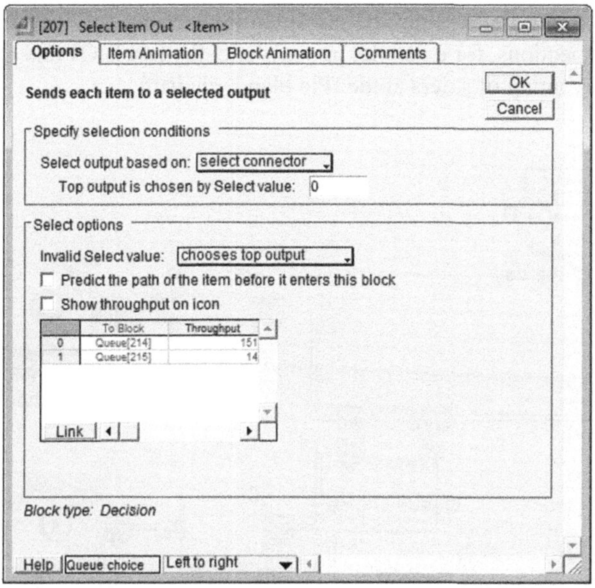

Figure 1.13. *The configured 'Queue choice' exit block selector (set to BY SELECT CONNECTOR)*

Our model should resemble the one shown in Figure 1.14.

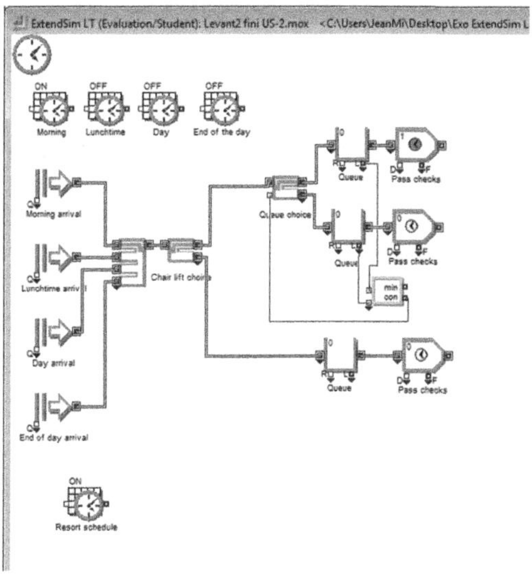

Figure 1.14. *Our model with the ski pass checks*

COMMENT 1.6.– In order to simplify and streamline the model, we can use labels to replace the connections, for example, in our model the labels Q1 and Q2 manage the queues and the routing of skiers at the 'Pic blanc' chairlift.

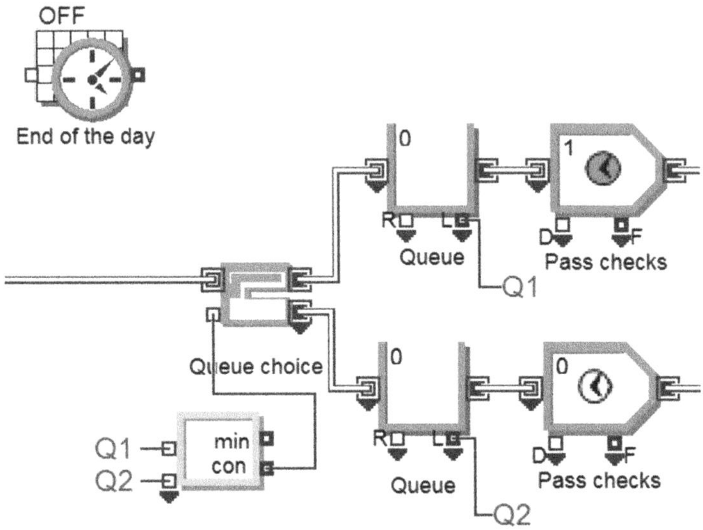

Figure 1.15. *The two labels Q1 and Q2 simulating the connections between the QUEUE blocks and the MIN & MAX block for the ski pass check at the 'Pic blanc' chairlift*

1.3.2.3. *Transporting the skiers*

Once their ski passes have been checked, the skiers queue in groups of four for 'Pic blanc' and groups of three for 'Plates', for the operator to assign them a chair. When they are sat down, the chairlift takes them to the drop-off point, where each skier chooses which piste to take.

We will start by forming batches of three or four skiers, representing a chair, *via* a BATCH block ('Item.lix' library) which we configure by inputting the QUANTITY NEEDED (Figure 1.16).

Behind each of these BATCH blocks, we will place an activity simulating the allocation of the chairs to the skiers. This will function according to the resort schedule, and each chair allocation has a constant duration of 18 s for the 'Pic blanc' chairlift and 15 s for the 'Plates' chairlift.

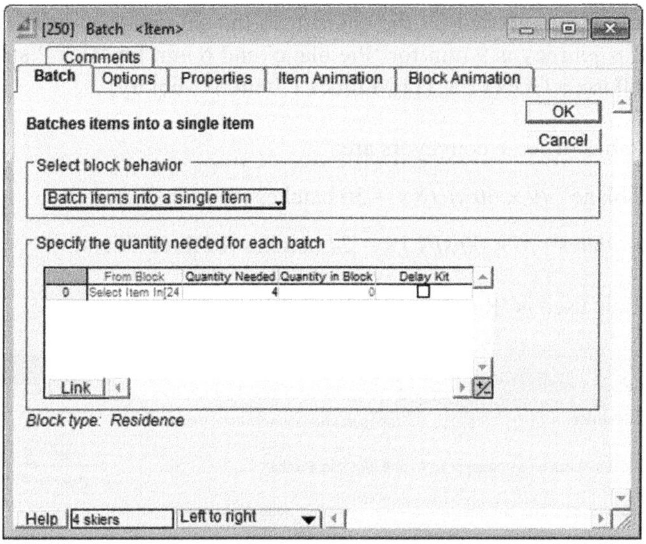

Figure 1.16. *Configuration of the BATCH block (quantity needed = 4) for four skiers ('Pic blanc')*

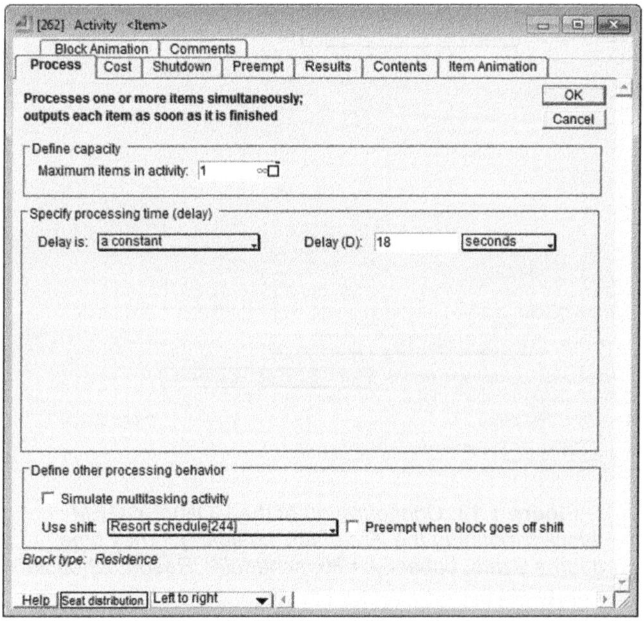

Figure 1.17. *Parameters for the chair supply blocks (schedule: 'Resort shift' and period: 18 s constant)*

The skiers are transported by the chairlift to the summit they have chosen. The duration of this journey is 9 min for 'Pic blanc' and 6 min for 'Des Plates'. For this action, we will use a CONVEY ITEM block ('Item.lix' library).

The capacities of these conveyors are:

– for 'Pic blanc': *(9 × 60 s)/18 s = 30* batches (of four skiers);

– for 'Des Plates': *(6 × 60 s)/15 s = 24* batches (of three skiers).

The schedule used is 'Resort schedule'.

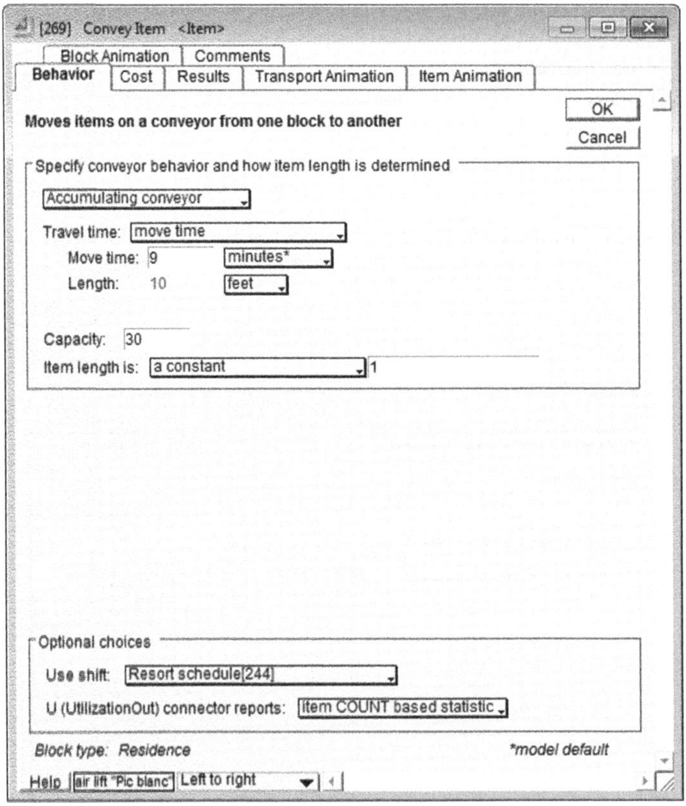

Figure 1.18. *Configuration of the CONVEY ITEM block simulating the 'Pic blanc' chairlift (journey time, duration = 9 min, capacity = 30, schedule: 'Resort schedule')*

When the chairlift arrives, the skiers must all be able to leave individually, so we split the batch (chair) using an UNBATCH block ('Item.lix' library).

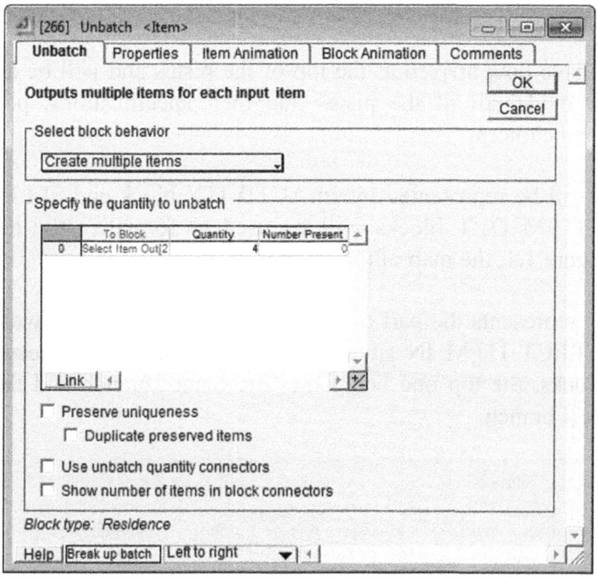

Figure 1.19. *Configuration for splitting up the batches on 'Pic blanc' with an UNBATCH block (quantity = 4)*

Our new model should resemble the one shown in Figure 1.20.

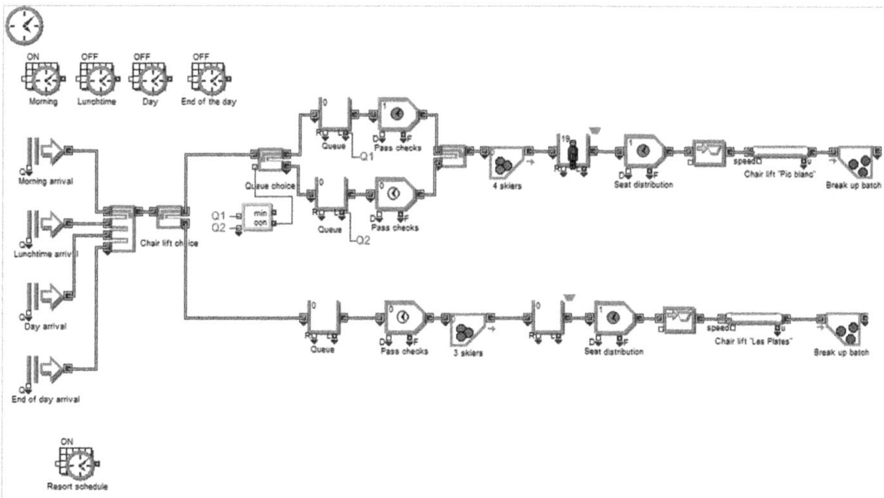

Figure 1.20. *Our model from the arrival of the skiers to the arrival of the chairlifts at the top of the pistes*

1.3.2.4. *Ski pistes*

The skiers have now arrived at the top of the pistes and will be able to ski, we must therefore model all of the pistes and their specifications: possible routes, descent times and choices.

Each piste will be represented by an ACTIVITY block and SELECT ITEM IN and SELECT ITEM OUT blocks will be used to formalize the available route choices (see Figure 1.1, the map of the resort).

Figure 1.22 represents the part of the model simulating the network of different pistes. The SELECT ITEM IN situated on the left will be connected to the two UNBATCH blocks, the top one being the 'Pic blanc' branch and the bottom one being the 'Plates' branch.

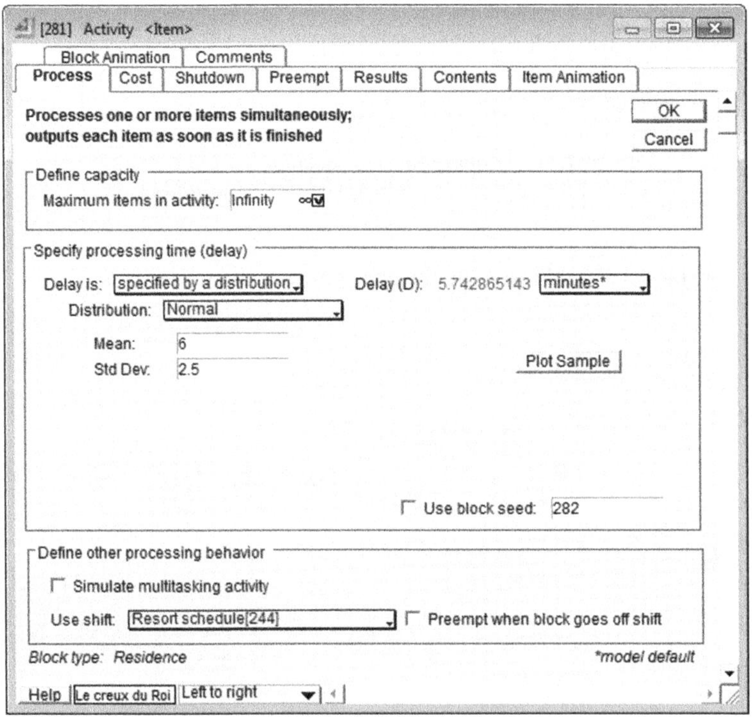

Figure 1.21. *An example of piste configuration, here 'Le creux du Roi' (Maximum number in the activity: ∞, Distribution: Normal, Mean: 6 min, Standard deviation: 2.5 min)*

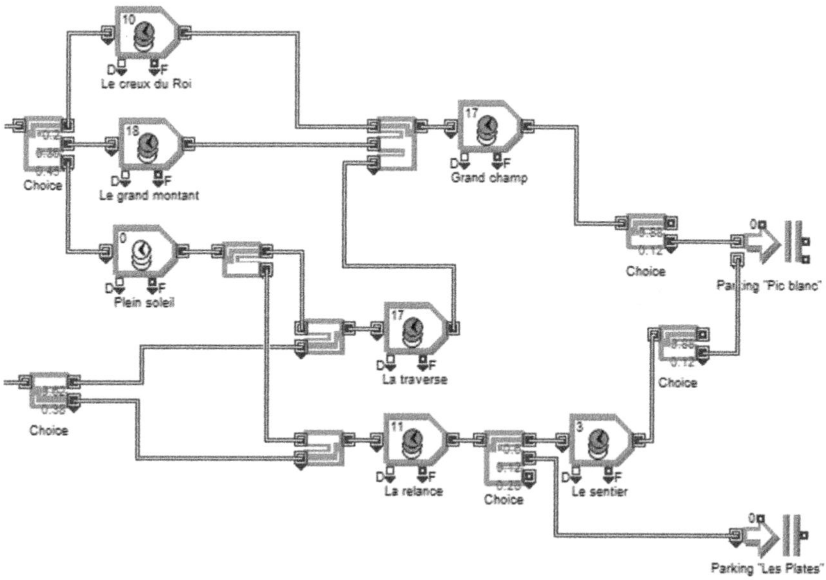

Figure 1.22. *The piste network, composed of ACTIVITY, SELECT ITEM IN and SELECT ITEM OUT blocks. Note: the probabilities are given on each of the EXIT SELECTOR blocks*

We will also include two EXIT blocks to represent the routes back to the parking lots for skiers deciding to stop skiing.

Two SELECT ITEM OUT blocks ('Item.lix' library) are placed before the exits allowing the skier to continue skiing by returning to each of the two chairlifts.

The probabilities for all of the selectors are those given in the design brief.

All of the pistes follow the 'Resort schedule' opening times.

Since the maximum number of skiers on the piste has not been specified, we assume that it is infinite.

To direct the skiers toward the chairlifts from the selectors situated behind the 'Grand champ', 'Le sentier' and 'La relance' pistes, we add, behind the 'Chairlift choice' selector, two input selectors – 'Towards Pic blanc' and 'Towards Les Plates' – as shown in 1.23, and we connect everything together.

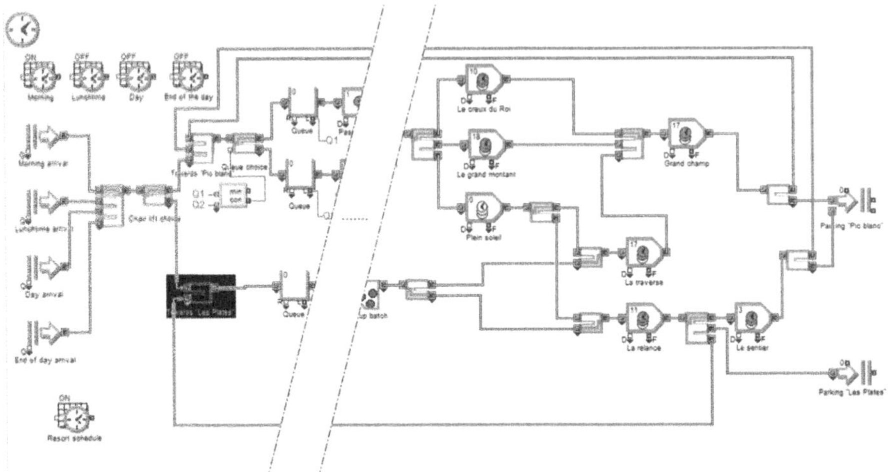

Figure 1.23. *The model with the routing of skiers toward the chairlifts via the SELECT ITEM IN blocks: toward 'Pic blanc' and toward 'Plates'*

1.3.3. *Definition of flows*

Having completed the design and its configuration following the design brief, we now have to define the flows which will be instigated by the 2D animation of our model.

For each of the four generators, we specify a type of item to animate, here we choose PERSON_M. The flows circulating around the model will be composed of these items, so we do not have to specify this elsewhere, they will automatically propagate through all of the blocks (see Figure 1.24).

1.3.4. *Running the simulation*

We can now run the simulation, after checking 2D ANIMATION and ADD CONNECTION LINE ANIMATION in the RUN menu. To do this, you simply have to click on the run icon found in the top toolbar.

The simulation can be controlled *via* the toolbar icons, whose functions are as follows:

– run the simulation: starts the simulation if all the parameters are correct;

– run Optimization or Scenarios: optimizes the simulation or controls the scenario if an OPTIMIZERor SCENARIO block is present in the model;

– stop: stops the simulation with no option to continue from this point later;

– pause/Continue: stops the simulation, but with the ability to restart from the point where it was stopped;

– step: carries out one step in the simulation followed by a pause;

– animation on/off: shows or hides the 2D animation while leaving the simulation to finish;

– animation Faster: accelerates the animation and reduces the overall duration of the simulation;

– animation Slower: slows the animation and increases the overall duration of the simulation.

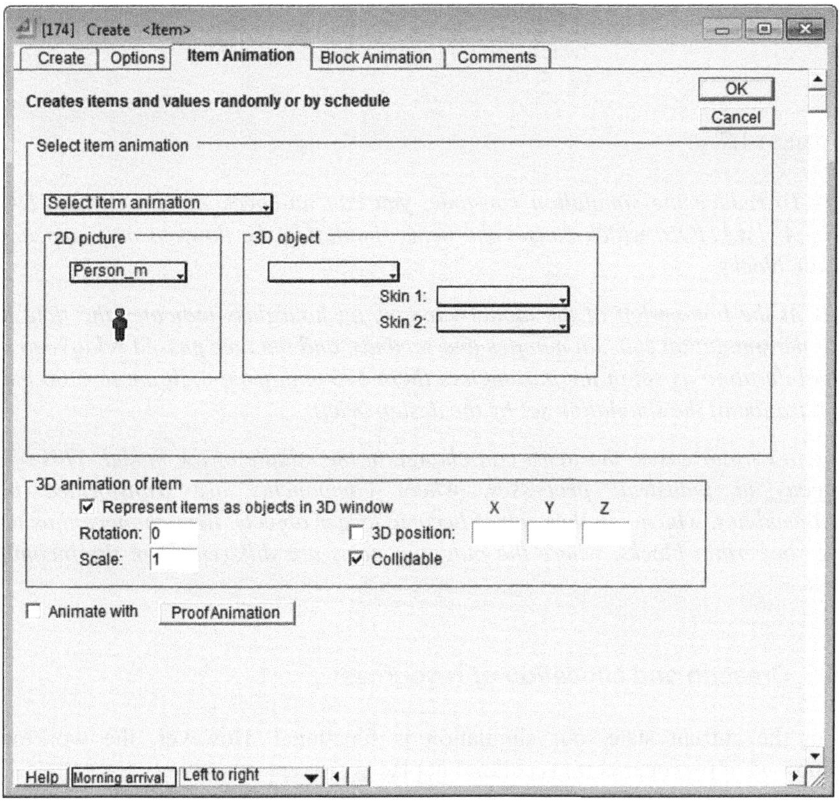

Figure 1.24. *The configuration of flows, here the transported items are of the PERSON_M type*

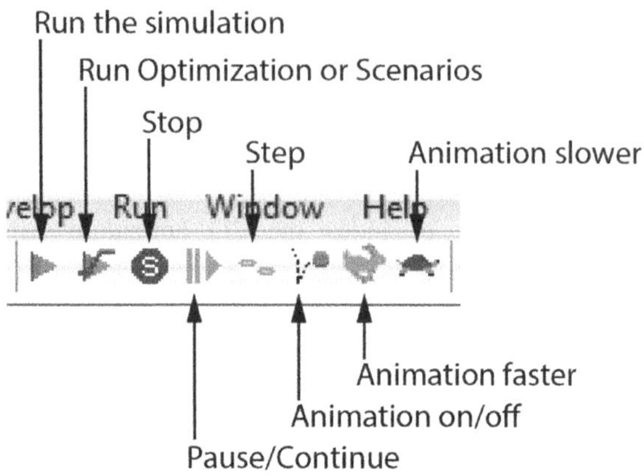

Figure 1.25. *The icons on the main toolbar which control the simulation run*

COMMENT 1.7.–

– To reduce the simulation run-time, you can un-check ADD CONNECTION LINE ANIMATION, which causes the items simulating the flows to only appear in certain blocks.

– At the bottom-left of the model window, an hourglass indicates the time left until the simulation ends, in minutes and seconds, and the time passed relative to the actual duration as set in the parameters (here 570 min, from 8:30 am to 6:00 pm – the duration of the simulation set by the design brief).

– In certain cases, the items can change in the middle of the model. This often happens in industrial processes, where components are transformed into subassemblies, which can themselves become larger objects. Item changes thus take place in certain blocks, where the outgoing items are different from the incoming items.

1.3.5. *Creation and allocation of resources*

In the current state, our simulation is functional. However, the workforce (composed of two operators) is still not in place.

The role of these operators, assigned to each ski lift, is to provide the chairs to the skiers whenever a row (a batch) of three or four skiers is formed.

To address this specification, we will have to create a RESOURCE POOL QUEUE to manage the workforce.

We place this RESOURCE POOL QUEUE block (QUEUE block, 'Item.lix' library) in front of our 'Seat distribution' ACTIVITY.

This ACTIVITY will be fed by a RESOURCE POOL block ('Item.lix' library) composed of two operators who work according to the opening times of the resort, from 9:00 am to 17:30 pm, at a cost of $18.25 per hour.

We insert a RESOURCE POOL block in our model and input the necessary parameters.

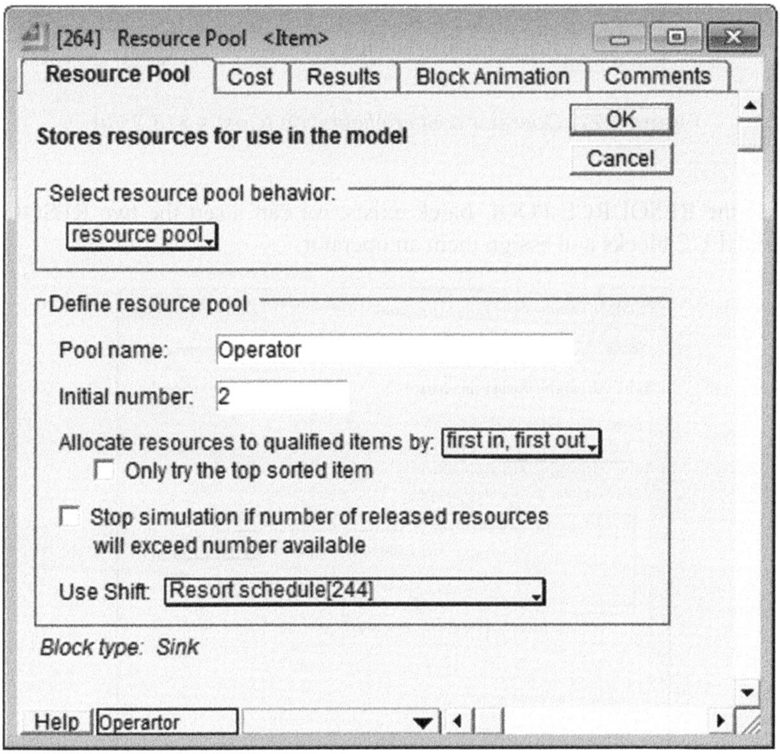

Figure 1.26. *Parameters for the 'Operators' RESOURCE POOL (Name of the pool: Operator, Initial number: 2, schedule: 'Resort schedule')*

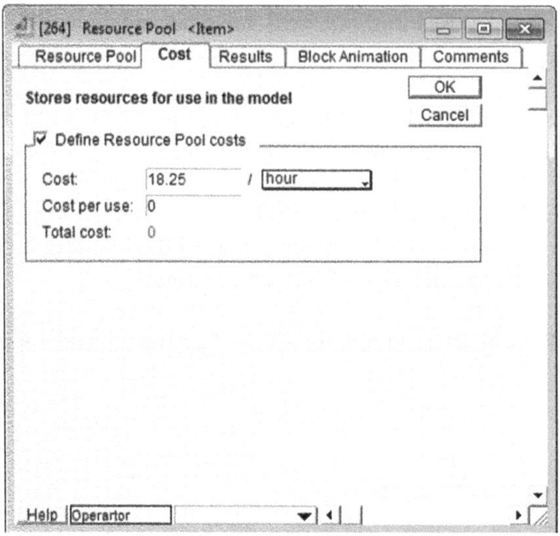

Figure 1.27. *Operator cost configuration (Cost = $18.25/h)*

Once the RESOURCE POOL block exists, we can insert the two RESOURCE POOL QUEUE blocks and assign them an operator.

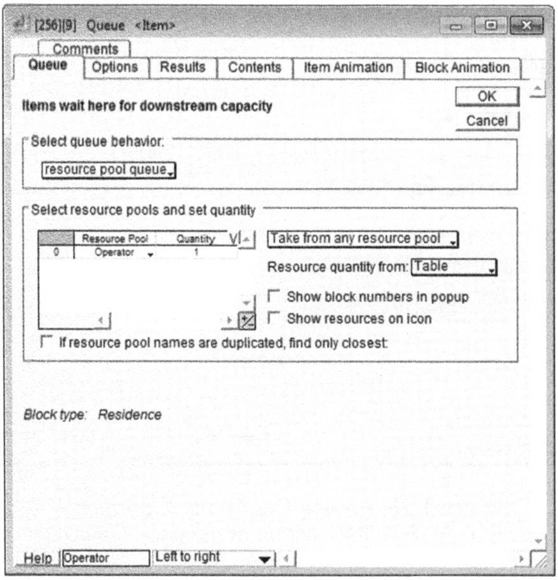

Figure 1.28. *Parameters for the QUEUE RESOURCE blocks for the operators (queue resource, Pool: Operator and Quantity = 1)*

Between each chair allocation, the operator must wait 15 or 18 s, or even longer if no skier is present (which is unlikely). We will thus free them of this work, while they wait, which is the role of the RESOURCE POOL RELEASE block ('Item.lix' library) (Figure 1.29).

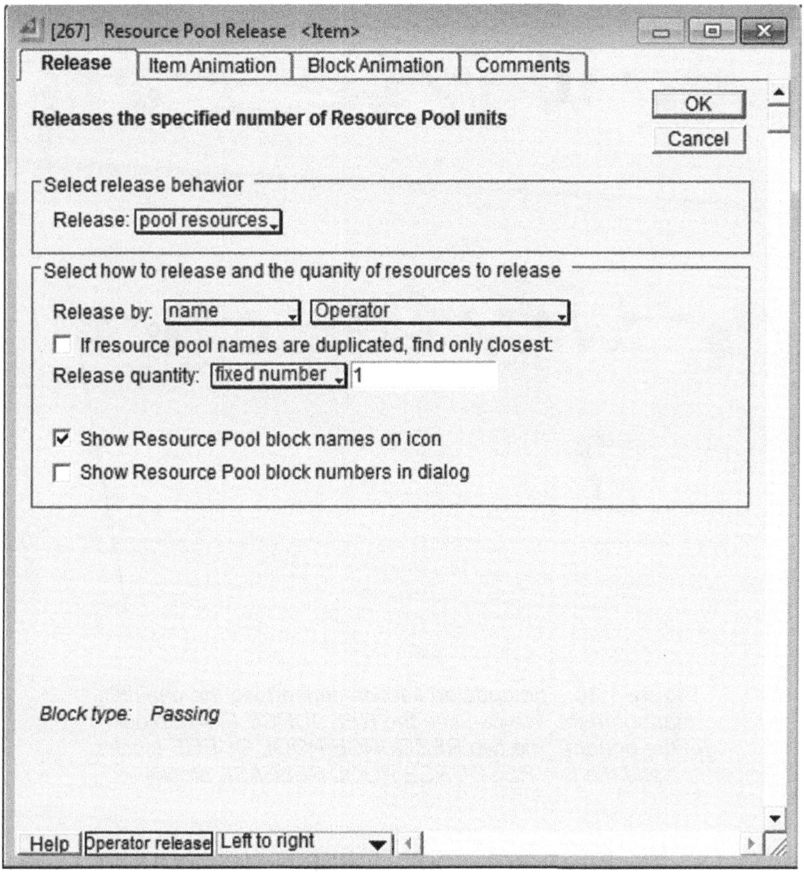

Figure 1.29. *Configuration of the RESOURCE POOL RELEASE block (name: operator, fixed number = 1)*

We place this block behind the 'Seat distribution' ACTIVITY block.

Our updated model, including the two operators, should resemble the one shown in Figure 1.30.

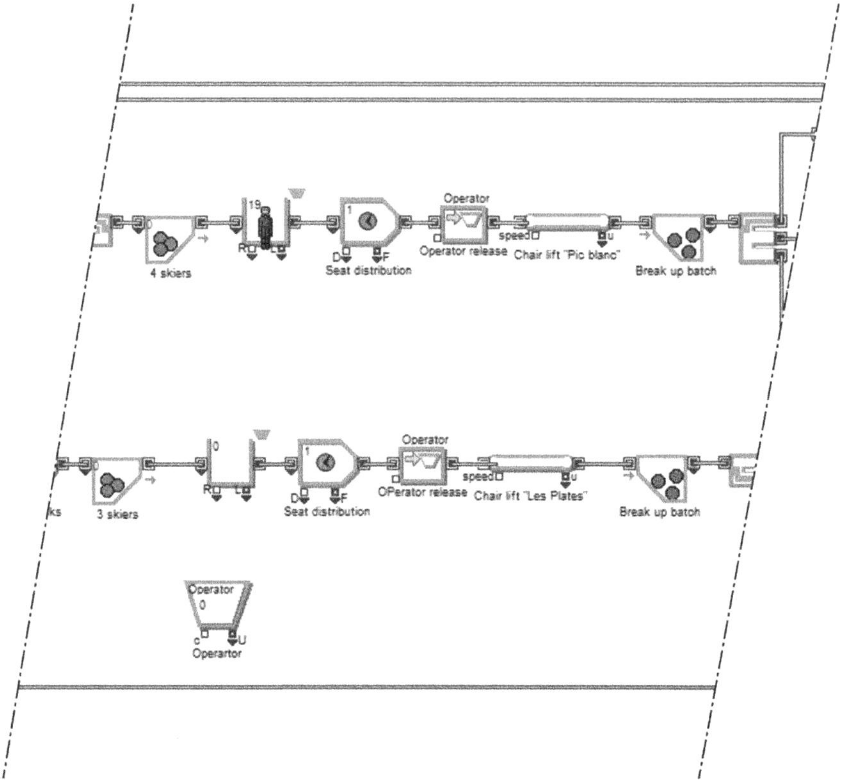

Figure 1.30. *The updated section concerning the chairlift management. We can see the RESOURCE POOL block (at the bottom), the two RESOURCE POOL QUEUE blocks and the two RESOURCE POOL RELEASE blocks*

COMMENT 1.8.– The RESOURCE POOL QUEUE block is a QUEUE block from the 'Item.lix' library, which we have configured to be in the QUEUE RESOURCE mode. To differentiate it, the icon representing it has a small, green trapezium above it.

1.3.6. *Rerunning the simulation*

We can rerun the simulation. Check that SHOW 2D ANIMATION and ADD CONNECTION LINE ANIMATION are checked in the RUN menu, otherwise no movement in the model will be visible.

1.3.7. *Generating a report and analysis*

Once our simulation model is operational, it is crucial to have an overall view of all the process data in order to check if certain specifications have been met and perhaps to make modifications.

It is the role of the report generator to amalgamate all of the results into a text-based report.

COMMENT 1.9.– For very complex models, the simulation can take a very long time. In order to reduce this time, we can set ExtendSim to conduct the whole simulation without any animation. In this case, the run-time is significantly shortened without affecting the calculation results or the values produced by the software. To carry out a simulation without animation, you must simply un-check 2D ANIMATION in the RUN menu before running the simulation.

To activate the report generator, the GENERATE REPORT option must be checked in the RUN menu.

Beforehand, we can choose the type of report generated from a list of three:

– dialogs (text file): gives all of the block input/output parameters and the information in the comments fields;

– statistics (text file): gives all of the final values of certain output parameters in a table format, ready for exporting as a table;

– statistics (DB): same as the previous option, but generates a database with tables. This database can be visualized and manipulated *via* the ExtendSim database management tool.

We are interested in the second type, 'Statistics (text file)', which will give us all of the important data from the simulation of our model in a coherent state.

In order to not overload our report, we will choose from all of the blocks only those whose calculated statistical results merit our attention.

To start, we select the 'Arrival' (CREATE), 'Queue' (QUEUE), 'Pass checks' (ACTIVITY), 'Chairlifts' (CONVEY ITEM), 'Pistes' (ACTIVITY), 'Operator' (RESOURCE POOL) and 'Parking lot' (EXIT) blocks. To do this, we hold down the CTRL button on the keyboard and click on each of these blocks.

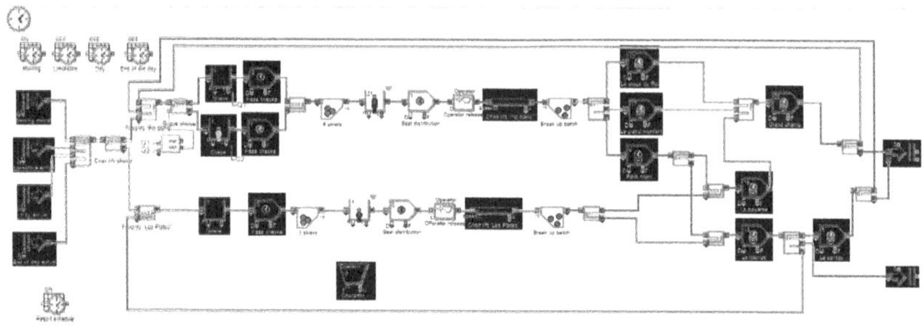

Figure 1.31. *Our model with the blocks selected for the report*

To include these blocks in our future report, select ADD SELECTED TO REPORT in the RUN menu.

When we now launch the simulation, a dialog box opens, asking us the name of the report ('report.txt' by default) and the folder we want to save it in.

By clicking on the SAVE button, the simulation runs and then the report is displayed in a text window and is saved.

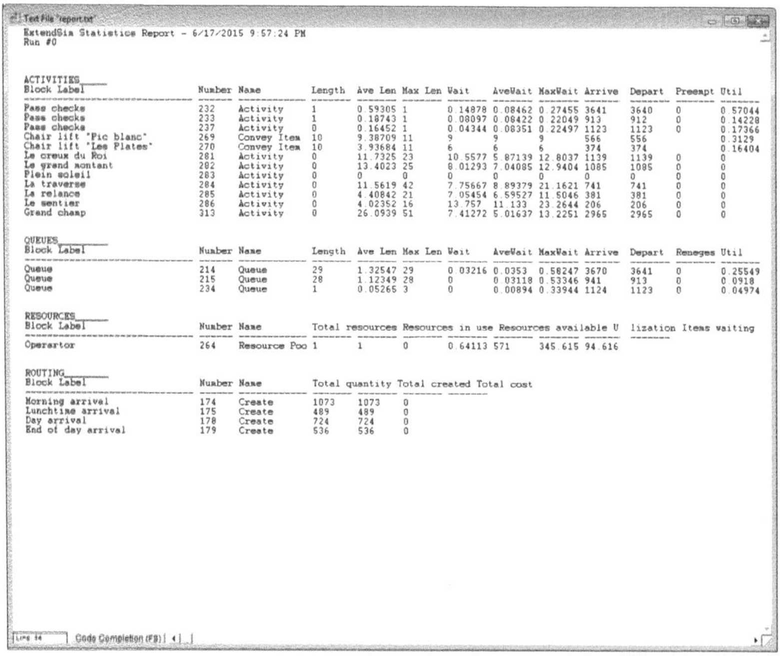

```
ExtendSim Statistics Report - 6/17/2015 9:57:24 PM
Run #0

ACTIVITIES
Block Label                 Number Name          Length Ave Len Max Len Wait    AveWait MaxWait Arrive Depart Preempt Util
Pass checks                 232    Activity      1      0.59305 1       0.14878 0.08462 0.27455 3641   3640   0       0.57044
Pass checks                 233    Activity      1      0.18743 1       0.08097 0.08422 0.22049 913    912    0       0.14228
Pass checks                 237    Activity      0      0.16452 1       0.04344 0.08351 0.22497 1123   1123   0       0.17366
Chair lift "Pic blanc"      269    Convey Item   10     9.38709 11      9       9       9       566    556            0.3129
Chair lift "Les Plates"     270    Convey Item   10     3.93684 11      6       6       6       374    374            0.16404
Le creux du Roi             281    Activity      0      11.7325 23      10.5577 5.87139 12.8037 1139   1139   0       0
Le grand montant            282    Activity      0      13.4023 25      8.01293 7.04085 12.9404 1085   1085   0       0
Plein soleil                283    Activity      0      0       0       0       0       0       0      0      0       0
La traverse                 284    Activity      0      11.5619 42      7.75667 8.89379 21.1621 741    741    0       0
La relance                  285    Activity      0      4.40842 21      7.05454 6.59527 11.5046 381    381    0       0
Le sentier                  286    Activity      0      4.02352 16      13.757  11.133  23.2644 206    206    0       0
Grand champ                 313    Activity      0      26.0939 51      7.41272 5.01637 13.2251 2965   2965   0       0

QUEUES
Block Label                 Number Name          Length Ave Len Max Len Wait    AveWait MaxWait Arrive Depart Reneges Util
Queue                       214    Queue         29     1.32547 29      0.03216 0.0353  0.58247 3670   3641   0       0.25549
Queue                       215    Queue         28     1.12349 28      0       0.03118 0.53346 941    913    0       0.0918
Queue                       234    Queue         1      0.05265 3       0       0.00894 0.33944 1124   1123   0       0.04974

RESOURCES
Block Label                 Number Name          Total resources Resources in use Resources available U lization Items waiting
Operartor                   264    Resource Poo  1      1       0       0.64113 571     345.615 94.616

ROUTING
Block Label                 Number Name          Total quantity Total created Total cost
Morning arrival             174    Create        1073   1073    0
Lunchtime arrival           175    Create        489    489     0
Day arrival                 178    Create        724    724     0
End of day arrival          179    Create        536    536     0
```

Figure 1.32. *The text window with our report in a table format*

COMMENT 1.10.–

The GENERATE REPORT option in the RUN menu remains checked if it has already been checked once and a new report is generated for each simulation run.

A dialog box asks you its name. If you do not change it, the old report will be overwritten. Hence, you must un-check this option if you do not want any further reports.

The format of the report stays the same unless it is deliberately changed.

Other report options are available in the RUN menu: 'REMOVE ALL FROM REPORT', 'ADD ALL TO REPORT', etc.

In the design brief (section 1.2.2) and slightly later on, where the queues are constructed (section 1.3.2.2), a maximum queue size of 25 skiers was specified.

Looking at the report, we can see that the maximum lengths of the three queues were 23, 14 and 22 skiers, respectively, which satisfies the specified constraint.

1.3.8. *Development, enhancement and improvement*

We will continue by improving the process involved in our model. To do this, new specifications as well as certain result indicators will be added.

1.3.8.1. *Categories of skiers*

After taking some surveys during the previous season, the sports department of the 'Levant' local authority confirms the following figures:

– 38% of skiers are men.

– 36% of skiers are women.

– 26% of skiers are children (under 16 years old).

To sort the skiers into the three categories (men, women and children), we will generate them randomly, as the skiers arrive.

We start by placing a RANDOM NUMBER block taken from the 'Value.lix' library, in our model, next to the four 'Arrival' ACTIVITY blocks as well as a SET block from the 'Item.lix' library.

We connect them together as shown in Figure 1.33.

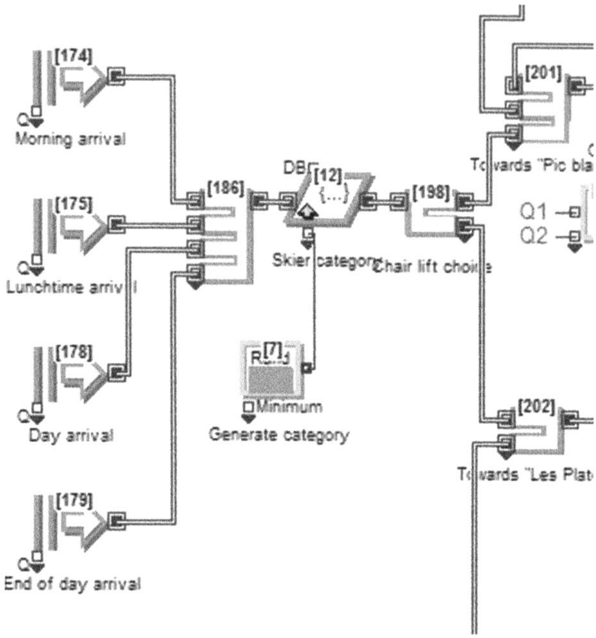

Figure 1.33. *In the center, the RANDOM NUMBER blocks ('Generate category') and SET ('Skier category') in our model*

COMMENT 1.11.– In Figure 1.33, the numbers of the blocks appear, which is an option given in ExtendSim – in the MODEL menu, check SHOW BLOCK NUMBERS.

Having opened the SET block, we choose NEW STRING ATTRIBUTE from the PROPERTY NAME column to create the skier categories.

A dialog box opens, in which we input 'CatSkier' (which will be the general variable governing the three categories of skiers), and then click on OK.

The EXECUTIVE block dialog box opens and we see our variable in the left-hand table, in the STRING ATTRIBUTE column.

We now input our three categories, 'man', 'woman' and 'child' into the CATSKIER column of the right-hand table, as shown in Figure 1.34.

We must again assign a different item to each of our categories in order to be able to visually differentiate them in the 2D animation.

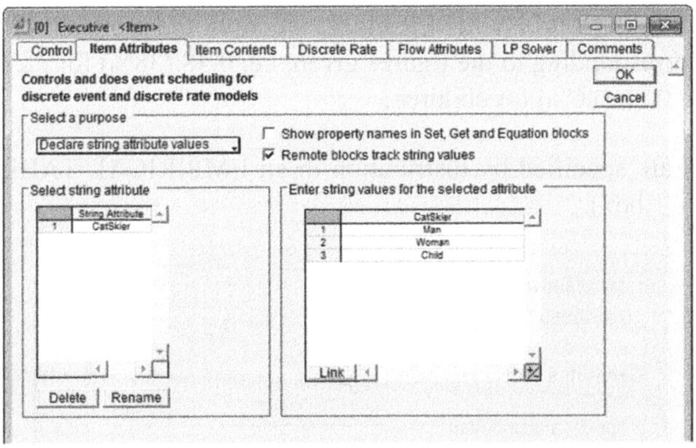

Figure 1.34. *The EXECUTIVE block dialog box with its two tables containing the network attribute variable, 'CatSkier' (left-hand table) and the three network value categories (right-hand table)*

We open the SET block and select the ITEM ANIMATION tab. Having chosen CHANGE ITEM ANIMATION USING PROPERTY and CatSkier in the two drop-down menus, we assign over three rows (green +/– box), the following items for each of the categories:

– PROPERTY VALUE: 'Man', 2D SYMBOL: 'Person_m';

– PROPERTY VALUE: 'Woman', 2D SYMBOL: 'Person_f';

– PROPERTY VALUE: 'Child', 2D SYMBOL: 'Person_labor'.

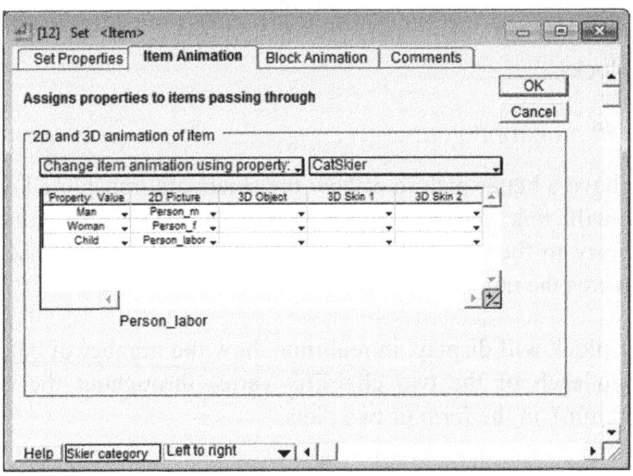

Figure 1.35. *Configuration of the SET block, with the categories (attributes) and their 2D symbols*

Next, in the RANDOM NUMBER block, we input the probabilities (in the form of decimals) corresponding to the figures given, i.e. 0.38 (38%) for men, 0.36 (36%) for women and 0.26 (26%) for children.

The values are specified by distribution in an EMPIRICAL TABLE with three values (green +/– box).

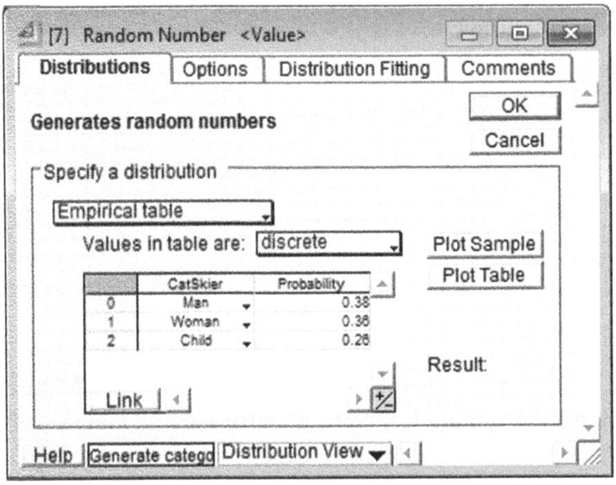

Figure 1.36. *Configuration of the probabilities of our categories in the RANDOM NUMBER block*

We can launch a simulation after checking SHOW 2D ANIMATION and ADD CONNECTION LINE ANIMATION in the RUN menu. The different items representing the different skier categories should appear on the connections and inside certain blocks.

1.3.8.2. *Chairlift monitoring curve*

In order to have a better picture of how the skiers are transported on each of the chairlifts, we will link the PLOTTER, DISCRETE EVENT block from the 'Plotter.lix' library to the output 'u' of each of the CONVEY ITEM blocks. This output 'u' measures the number of items conveyed.

This plotter block will display, in real-time, how the number of batches of skiers who go through each of the two chairlifts varies throughout the course of the simulation (570 min), in the form of two plots.

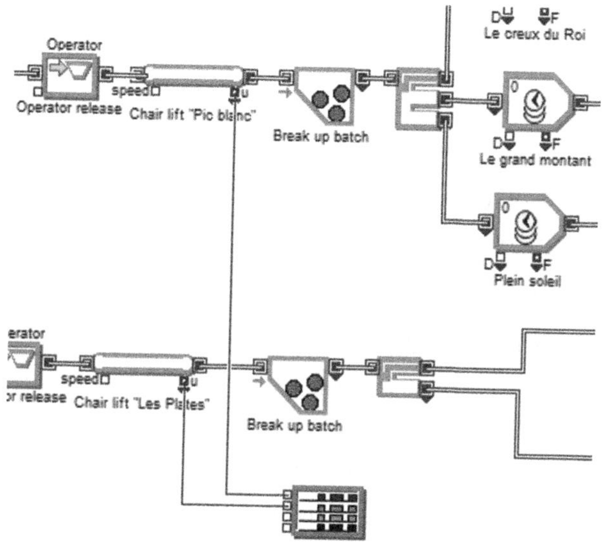

Figure 1.37. *The PLOTTER, DISCRETE EVENT block connected to the outputs 'u' of the CONVEY ITEM blocks (chairlifts)*

We still need to configure this block.

When we open the block, we find a toolbar from which we can define the necessary configuration parameters.

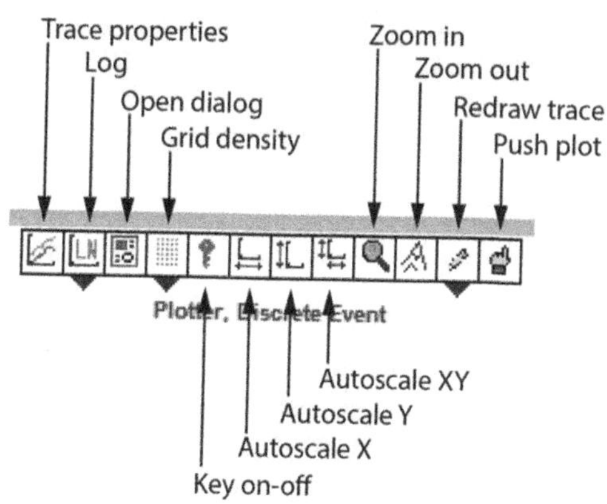

Figure 1.38. *The toolbar for PLOTTER blocks*

We click on the icon that is furthest to the left (the ‘Trace properties’ tool). A table with nine columns appears in which we can modify, from left to right:

– text describing the trace;

– the trace color (seven colors);

– the trace width (five widths);

– the trace pattern (four patterns);

– the trace style (interpolated, stepped or points);

– the trace symbols (dots, squares, + signs, circles, etc.);

– the number format for the data pane (general: Gen, decimal: x.xx, integer: xxx and scientific notation: x.xe);

– the plot axis for the trace (the left-hand axis, Y1, or the right-hand axis, Y2);

– the visibility (visible: open eye or invisible: closed eye).

We fill in the table as shown in Figure 1.39.

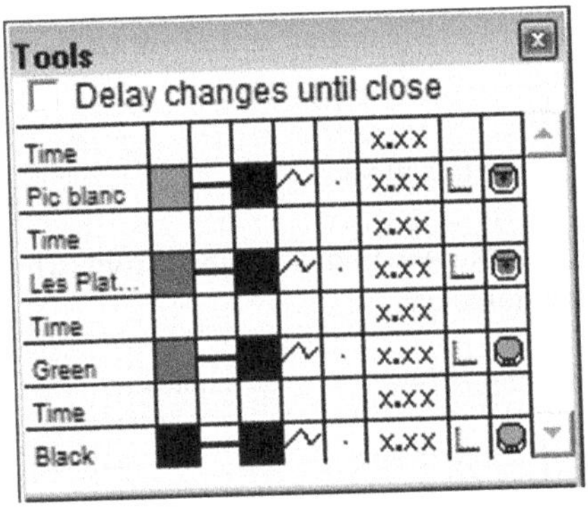

Figure 1.39. *The parameters to be input and chosen in the trace properties window. For a color version of this figure, see www.iste.co.uk/reveillac/modeling3.zip*

We then select the third tool from the left (the ‘Open dialog’ tool) and configure it by checking SHOW PLOT DURING SIMULATION (the plot will be traced in real-time during the simulation), SHOW INSTANTANEOUS QUEUE LENGTH (the queue length will be visible in zero time), AUTOSCALE DURING SIMULATION (the plot scale will be recalculated in real-time as new data arrive) in the drop-down menu.

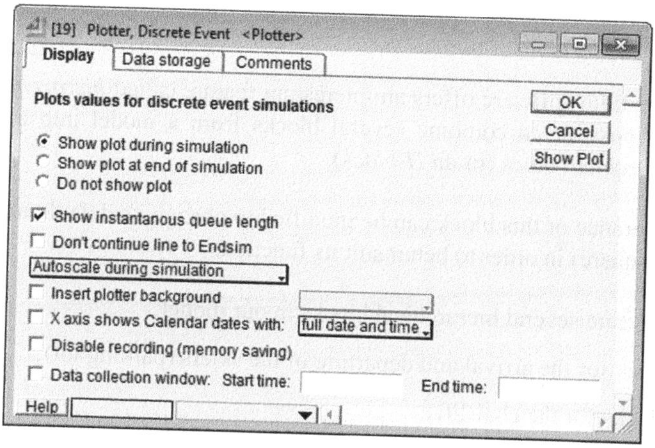

Figure 1.40. *Configuration of the dialog tool*

We can rerun the simulation, the plotter window will open and the trace plots itself in real-time throughout the simulation.

A table situated underneath the plot shows the key figures (time and number of items/batches) for each point of the trace for each chairlift.

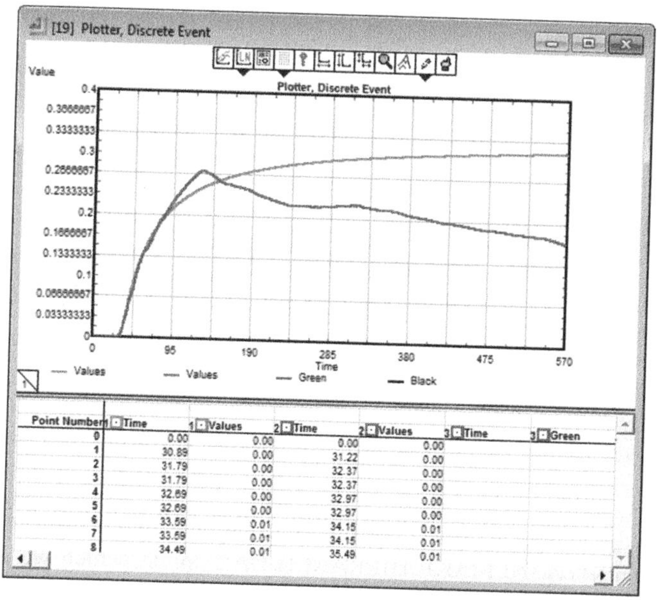

Figure 1.41. *The plot display window with its data-tracking table. For a color version of this figure, see www.iste.co.uk/reveillac/modeling3.zip*

1.3.9. *Hierarchy*

The ExtendSim software offers an interesting feature called *hierarchy*. With this feature, it is possible to combine several blocks from a model into a new block called a *hierarchical block* (or an *H-block*).

The appearance of this block can be modified to change its default appearance (a simple gray square) in order to better suit its function.

We will create several hierarchical blocks in our model:

– one block for the arrival and departure of the skiers (parking lot);

– two blocks for the chairlifts;

– seven blocks for the pistes.

A hierarchical block takes all of the inputs and outputs already present in the model. In Figure 1.42, we can see the three inputs, related to the 'Pic blanc' ski pass check, the three outputs, related to the pistes and the output monitoring the number of skiers, connected to the PLOTTER, DISCRETE EVENT block.

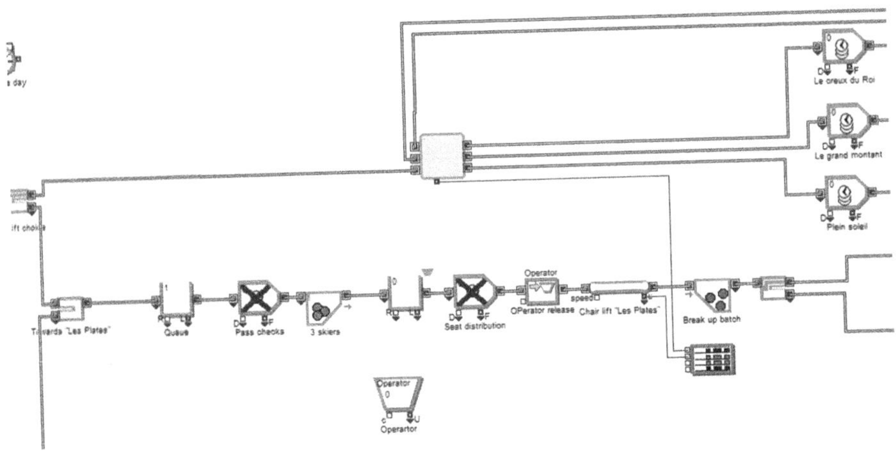

Figure 1.42. *The H-block (in the middle) which substitutes for all of the 'Pic blanc' chairlift blocks from the queues to the skiers' arrival at the pistes*

To build this block, we select the blocks to be built into a hierarchy and then right-click and select the MAKE HIERARCHICAL option, which we name in the dialog box that opens (Figure 1.43).

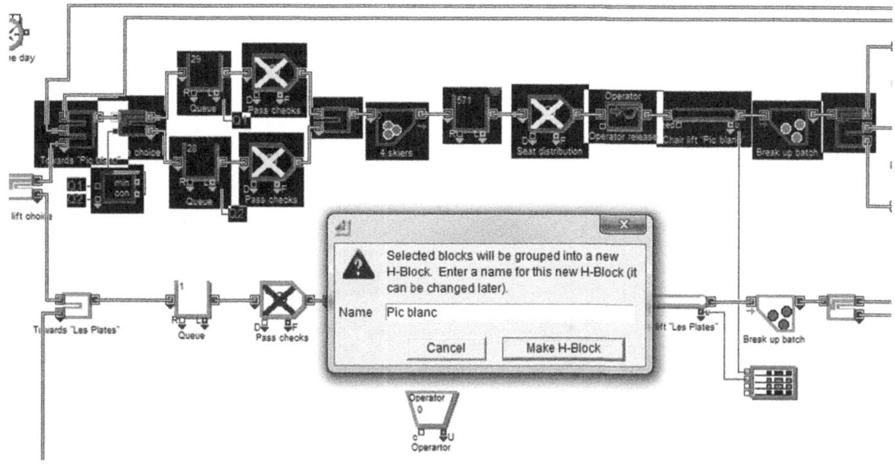

Figure 1.43. *Creating the 'Pic blanc' H-block*

To open the H-block, we can right-click on it and select OPEN STRUCTURE. In the window that opens (which is divided into four zones), we can make modifications, reorganize the inputs/outputs, change the appearance of the block, change the connections, etc.

The full variety of available features goes far beyond the scope of this chapter, and they will not all be described here.

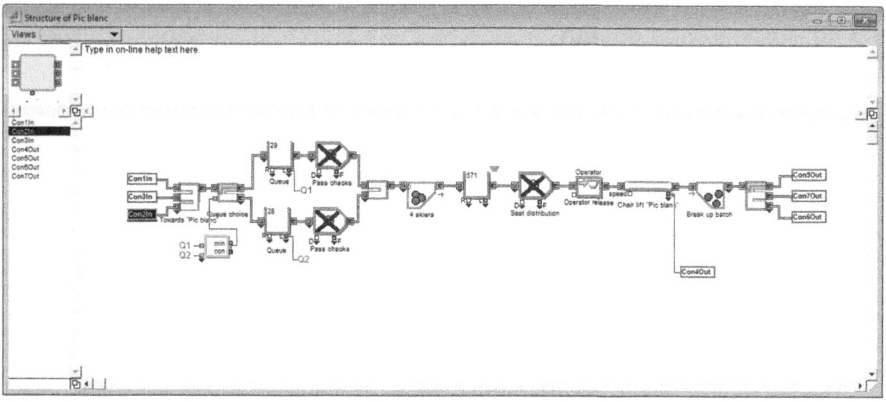

Figure 1.44. *The reorganized 'Pic blanc' H-block structure. We can see the graphical representation (at top-left), the list of inputs and outputs (bottom-left), the help text (top-right) and the submodel of the H-block with its internal inputs/outputs*

By applying this hierarchical technique, we can create all of the H-blocks proposed above, and change their appearance to end up with the model shown in Figure 1.45.

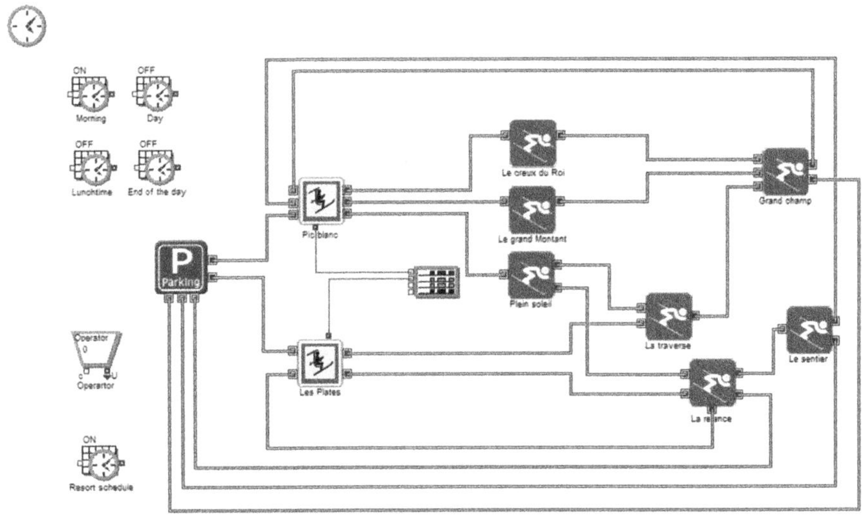

Figure 1.45. *Our model with all of the H-blocks. The gray squares that represent each of the blocks by default have been replaced by logos*

1.3.10. *Appearance design*

It is possible to greatly improve the interface of a 2D model by adding text, button commands, menus, dialog zones, etc.

Several tools exist to enable this:

– text, image and logo management;

– customized function creation from blocks of the 'Utilities.lix' library;

– configuration or attribute cloning from the dialog boxes.

The top toolbar has icons offering a variety of graphical functions, some linked to particular menus.

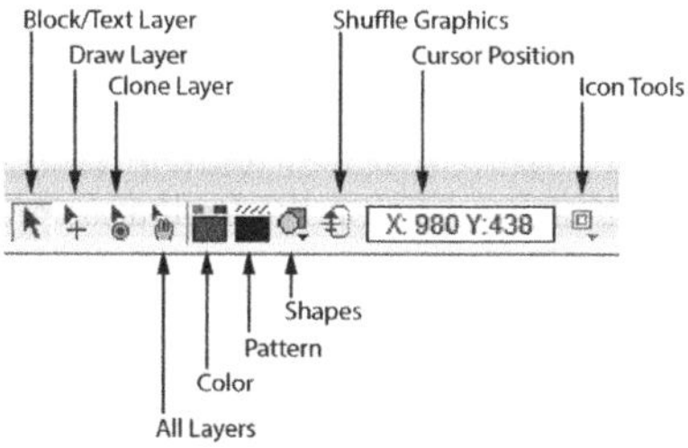

Figure 1.46. *Toolbar icons that can be used changing the design, among other things*

We will create:

– a title linked to a logo;

– a button for running the simulation;

– a button for pausing the simulation;

– two interactive dialog zones for configuring the transport time for each ski lift and the arrival frequency of the chairs.

We start by inputting the text for the title and the two dialog zones (double-click at the bottom of the model and use the TEXT menu), as in Figure 1.47. The two boxes are made using the 'Graphics layer' tool, and their color can be set with the 'Colors' tool. The logo is put in place by copying and pasting a preselected jpeg image.

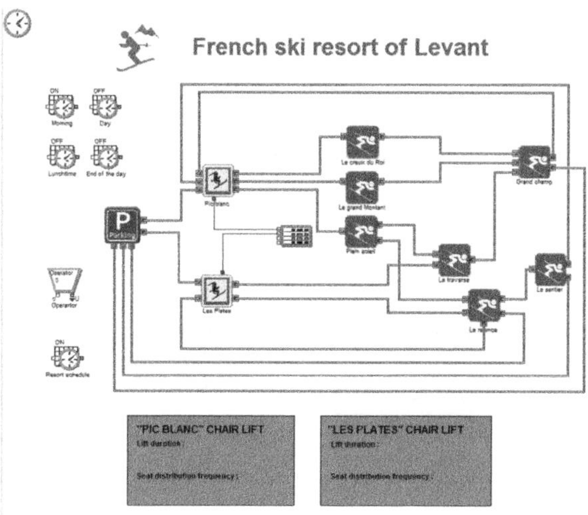

Figure 1.47. *The design elements: title, logo, dialog zones and boxes*

We clone the '9 minutes* duration' and '6 minutes* duration' attributes, the two 'Chairlift' CONVEY ITEM blocks and the '18 second period (P)' and '15 second period (P)' attributes from the two 'Seat distribution' ACTIVITY blocks into each of our zones using the 'Clone layer' icon on the toolbar.

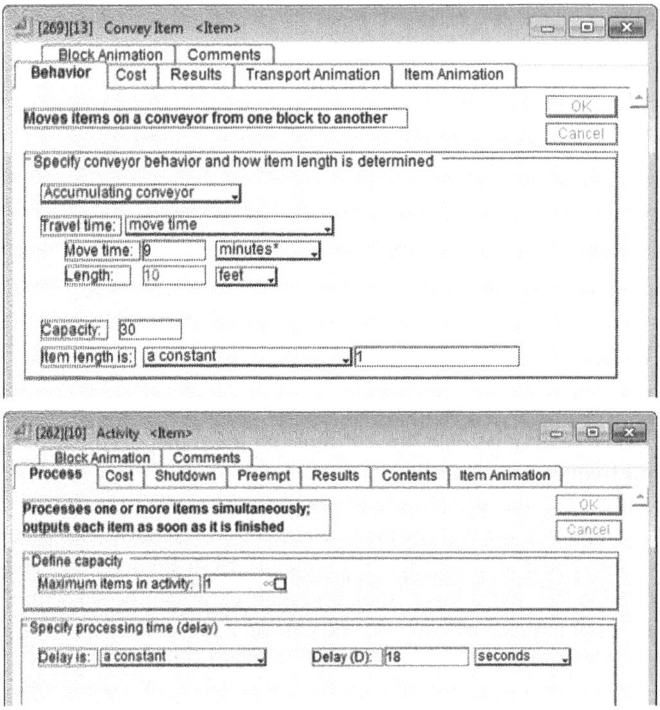

Figure 1.48. *The 'Move time ... minutes*' attributes to clone into the 'Chairlifts' CONVEY ITEM blocks (above) and the (see figure 1.48 below): 'Delay (D) ... seconds' attributes to clone into the 'Seat distribution' ACTIVITY (below)*

We should finally obtain dialog zones identical to the ones shown in Figure 1.49.

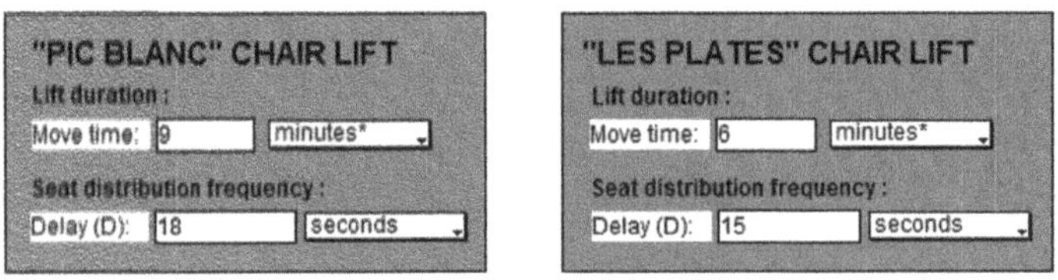

Figure 1.49. *The two finished dialog zones*

To create the buttons, we place two BUTTON blocks from the 'Utilities.lix' library in our model.

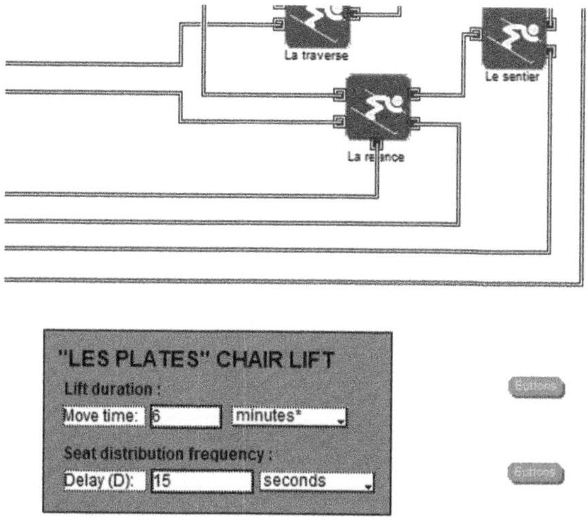

Figure 1.50. *At bottom-right, our two BUTTON blocks*

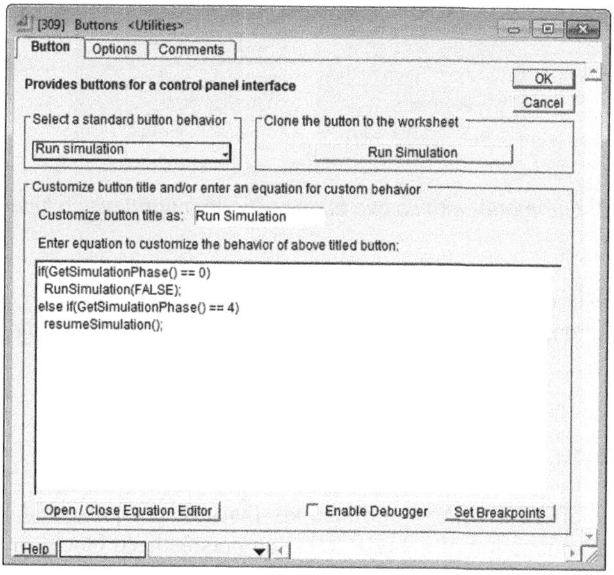

Figure 1.51. *The configuration of the 'Run Simulation' button, with the choice of the button's action and the cloning zone*

We configure the first with the RUN ANIMATION action and the second with the PAUSE SIMULATION action, then we clone each of the buttons (using the 'Clone layer' tool) into our model, taking care to place the clones over the BUTTON blocks to hide them.

Once finished, our model should resemble the one shown in Figure 1.52.

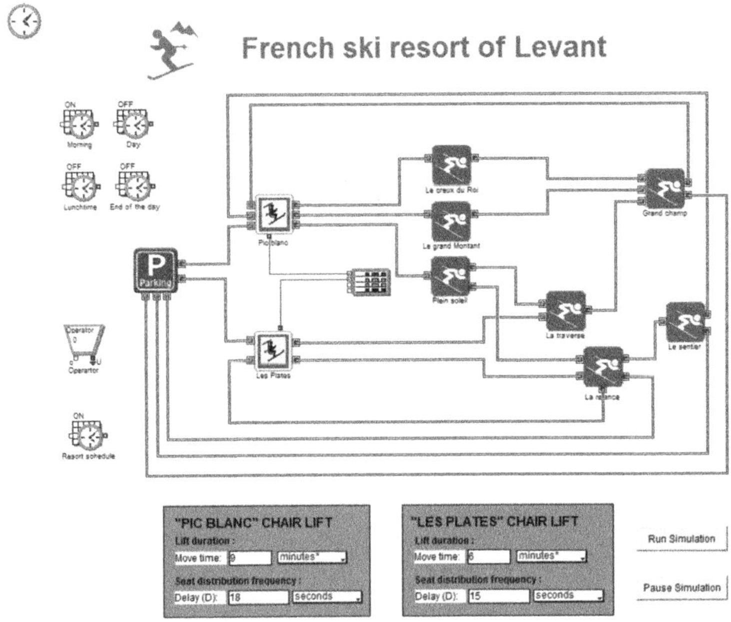

Figure 1.52. *Our model with its two buttons (bottom right) which hide the blocks*

Our model is now finished and we can run the simulation by clicking on the appropriate button, change the parameters of each of the chairlifts or pause the simulation.

1.4. Conclusion

ExtendSim offers a variety of functions that are not presented here, such as generating a 3D animation of a model, creating personalized blocks or manipulating databases using the Microsoft Excel *add-in*.

If you are interested and want to go beyond what has been presented in this chapter, you can study the software help and work through some of the examples provided.

A demonstration version of the ExtendSim software is available for download from the publisher's Website (see the Internet links at the end of the book).

Other, equally powerful, software exists, with several variations in use, functionalities and usability (see section 10.4 of Volume 1 of this book).

2

Simulation of Mixed Flows

2.1. Mixed Flows

In Chapter 1, we were able to apply computerized flow simulation to an example concerning the 'Levant' winter sports resort.

This simulation was based upon flows of skiers, which are countable, meaning that we were modeling *discrete* elements.

In this chapter, we will combine discrete and *continuous flows*, giving rise to the term *mixed flows*.

Continuous flows mostly arise in examples involving flows of liquid materials, for example water, molten metal or glass, oil, milk, gas, air or petroleum, and related to processes or industrial techniques such as melting, rolling or molding of various materials, food processing, piping or draining of rivers and canals, air flow in aerodynamics, processing and refining petroleum products or preparing chemical or pharmaceutical products.

It should be noted that processes based entirely upon continuous flows are rare. Much more usual are processes in which continuous flows are transformed into discrete flows, usually related to specific packaging types (cardboard boxes, cans, jars, components of larger products, etc.) during their progression through the process.

The example discussed in the following sections is again constructed using the ExtendSim 9 software.

COMMENT 2.1.–

Within the ExtendSim framework, continuous flow modeling makes use of several libraries:

– the 'Item.lix' library designed for discrete flows;

– the 'Rate.lix' library designed for continuous flows, even though some of its blocks can define discrete entities by, for example, changing the units (with the 'Change Units' block);

– exchanges between discrete and continuous entities are possible (using the 'Interchange'block).

2.2. An example of modeling mixed flows

For our mixed flow example, we will model a production and packaging line for fizzy drinks.

A fizzy drink is a mixture of water, artificial flavorings and preservatives, and a fruit mix with added sugar. After being mixed, it is bottled and carbonated before being labeled and packaged.

Two packaging types are available: crates, for professional customers, and packs, for private consumers.

The crates are simply stored in piles and stocked in a warehouse, whereas the packs are packaged before being stored in the warehouse.

The primary aim of this simulation is to determine the supply frequencies and rates of the different consumables used on the line. The production constraints of the product (soda) are defined according to a recipe and a predefined process.

2.2.1. *Problem statement and specifications*

Factory opening times: 6:00 am to 8:00 pm, Monday to Friday.

Production line timetable:

– production stage (operational): 12 h/day, from 8 am to 8 pm;

– preventative maintenance stage: 2 h/day, from 6 am to 8 am;

– cleaning stage (washing and rinsing): 1 h/day, from 7 pm to 8 pm.

2.2.1.1. *Production phase*

The water, artificial flavorings and fruit mix are stocked in independent, continuously supplied tanks. The throughput is measured in liters/minute (l/min).

Tank throughputs:

– water: 80 l/min;

– artificial flavorings: 1 l/min;

– fruit mix: 19 l/min.

COMMENT 2.2.– The specified flow rates determine the soda recipe, which is 80% water for 1% flavorings/preservatives and 19% fruit mix. These proportions must not change in order to obtain a drink of constant quality.

The mixture of the three products is then gradually heated and circulated around a heating conduit.

Heating:

– duration: 10 min;

– length: 10 m;

– maximum throughput: 200 l/min.

Once the mixture is at the correct temperature, the sugar (which is also stocked in a continuously supplied tank) is added.

Finishing the un-bottled soda:

– distribution throughput: 10 kg/min;

– weight-volume conversion factor for the sugar: 0.8 kg/l;

– percentage of sugar in the soda: 20%.

The resulting un-bottled soda is stored in a buffer tank with a maximum capacity of 2,000 l before being bottled.

COMMENT 2.3.–

– The bottling time is variable and depends upon the quantity of un-bottled soda in the upstream buffer tank, of maximum capacity 2,000 l.

– The quantity available in the tank controls the bottling rate.

– The buffer tank must be monitored to ensure it never empties or overflows.

– It is also important to take accumulation of the mixture in the heating conduit into consideration, as this places important constraints on throughput and run-time.

Once the bottles are full, they are taken by roller conveyor to the carbonation machine to be injected with carbon dioxide and produce the desired fizzy drink.

Bottling, transport and carbonation parameters:

– bottle capacity: 1.5 l;

– bottling duration (for each 1.5 l bottle): 1–5 s;

– transfer time (transport to carbonation machine): 20 s;

– roller conveyor capacity: 100 bottles;

– mean bottle diameter: 10 cm;

– carbonation time: 0.5 s for a batch of 1–6 bottles.

Following the carbonation, the bottle is labeled before being taken to two packaging lines, again by roller conveyor. The first line assembles and wraps packs of bottles in plastic film for mass-distribution, and the second line puts the bottles in plastic crates for professional use. Next, the crates are transported to a specific storage site and the packs are stored on pallets in specially equipped warehouses.

Labeling parameters:

– duration: 1 s per bottle;

– label supply: 1 label/1.5 s.

Packaging, transport and storage parameters (crates):

– crate capacity: 12 bottles;

– time taken to fill each crate: 6 s for 1 batch of 12 bottles;

– roller conveyor length: 30 m;

– roller conveyor speed: 5 m/s;

– roller conveyor capacity: 64 crates;

– time taken to put in storage: 30 s/crate.

Packaging, transport and storage parameters (packs):

– number of bottles per pack: 6;

– time taken to wrap in plastic film: 1.2 s/pack;

– roller conveyor length: 20 m;

– roller conveyor speed: 4 m/s;

– roller conveyor capacity: 40 packs;

– number of packs per pallet: 72;

– time taken to place on pallet: 20 s;

– time taken to store in warehouse: 4 min/pallet.

Only 15% of the bottles produced are put in crates, the other 85% are put in packs for mass-distribution.

2.2.1.2. *Maintenance and cleaning stages*

Preventative maintenance is completed each morning by a team of three technicians who check all of the machines on the production line. They replace worn components, carry out any required adjustments and complete the necessary upkeep.

The cleaning, which consists of emptying, cleaning and rinsing the tanks as well as all the piping carrying the raw ingredients, the blends and the soda, is completed each night by a team of five specialist workers.

Technicians' shift: 6:00 am–1:00 pm every day.

Specialist workers' shift: 1:00 pm–8:00 pm every day.

Outside of the preventative maintenance and cleaning periods, the general personnel, technicians and workers attend to other business.

COMMENT 2.4.– In reality, the technicians and workers could intervene in the event of unexpected breakdown on the line during the production period (from 8:00 am to 7:00 pm). This eventuality is not taken into consideration in this simulation.

2.2.1.3. *Overall simulation conditions*

The simulation takes place between 6:00 am and 8:00 pm on the 1st March 2016, i.e. one full working day of the factory.

This simulation will be based purely upon operational considerations; no costs are taken into account.

2.3. Creating and inputting the project in ExtendSim

As in Chapter 1, for pedagogical purposes, the project will be divided into several stages that we will put together as we construct the model:

– definition of the principal parameters;

– production of un-bottled soda followed by its bottling;

– transport, carbonation and labeling;

– packaging and storage;

– maintenance and cleaning;

– finishing touches.

2.3.1. *Definition of the principal parameters*

We first define a new model: FILE menu, NEW MODEL.

We now define the time parameters of the simulation: RUN menu, SIMULATION SETUP.

In the tab labeled DISCRETE, we check USE CALENDAR DATES and select 01/03/2016 for the START, as defined in the problem statement, and 06:00:00 as the start time of the simulation.

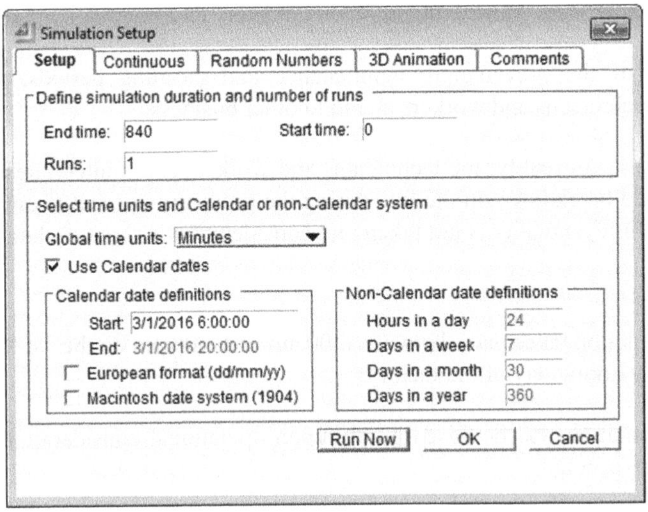

Figure 2.1. *The configuration of our simulation*

The default time unit will be minutes (GLOBAL TIME UNITS) as it is well suited to the future applications of the simulation.

The END TIME is set to 840 min, or 840/60 = 14 h, from 6:00 am to 8:00 pm.

In order to incorporate the parameters into the simulation, we place the EXECUTIVE block from the 'Item.lix' library in our model, in the top left-hand corner.

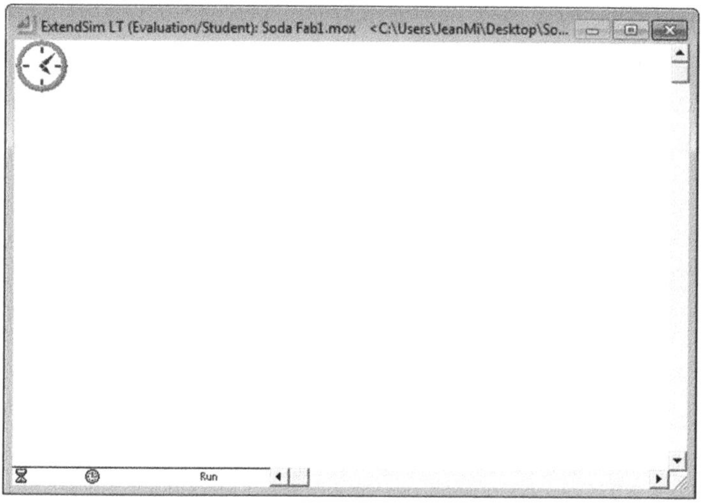

Figure 2.2. *The new model with its EXECUTIVE block*

2.3.2. *Soda production and bottling*

First, we focus our attention on the production of un-bottled soda, prior to bottling and carbonation.

At this stage, most of the flows are continuous. They become discrete upon bottling.

2.3.2.1. *Constructing and configuring the model*

We start by placing a SHIFT block ('Item.lix' library) in our model, by dragging it into the model.

By double-clicking on the block, we can define the working schedule of the production line, which we will call "Operational" (Figure 2.3).

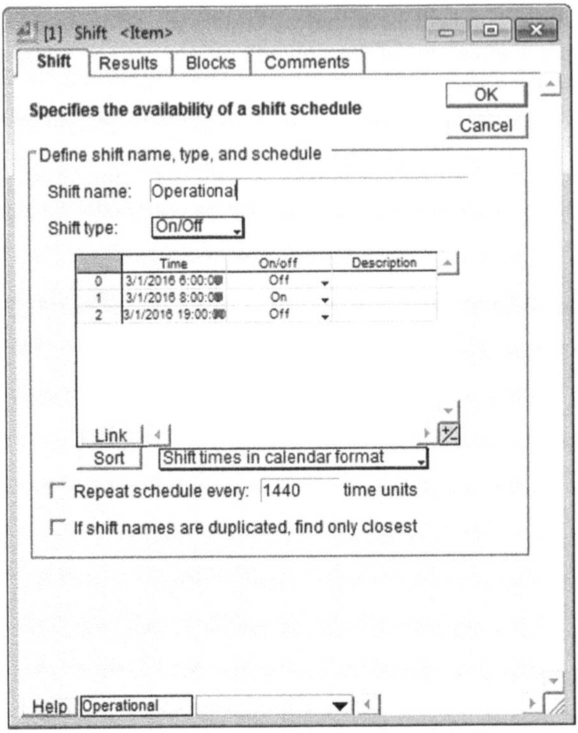

Figure 2.3. *The configuration window of the "Operational" SHIFT block*

In Figure 2.3, we can see the name and the schedule periods, with a start time of 6:00 – OFF, corresponding to the start time of our simulation that we previously set in the SIMULATION SETUP window (section 2.3.1), a time operations start, 8:00 – ON and a time operations stop, 19:00 – OFF.

The schedule is in calendar format as chosen in the SIMULATION SETUP.

COMMENT 2.5.–

– A calendar must always have a start time equal to the start time input in the simulation parameters.

– To add or delete a row, you simply click on the green +/– box.

– The ON position indicates the start of an active schedule period and the OFF position indicates the end of one.

– The schedule can be in numerical format, in which case we set a number of time units, using the units selected in SIMULATION SETUP, passed since 0 (the time the simulation started).

We will then add three TANK blocks ('Rate.lix' library) for the water, the artificial flavorings/preservatives and the fruit mix (Figure 2.4), which will be connected to three VALVE blocks ("Rate.lix" library).

We should obtain the model shown in Figure 2.4.

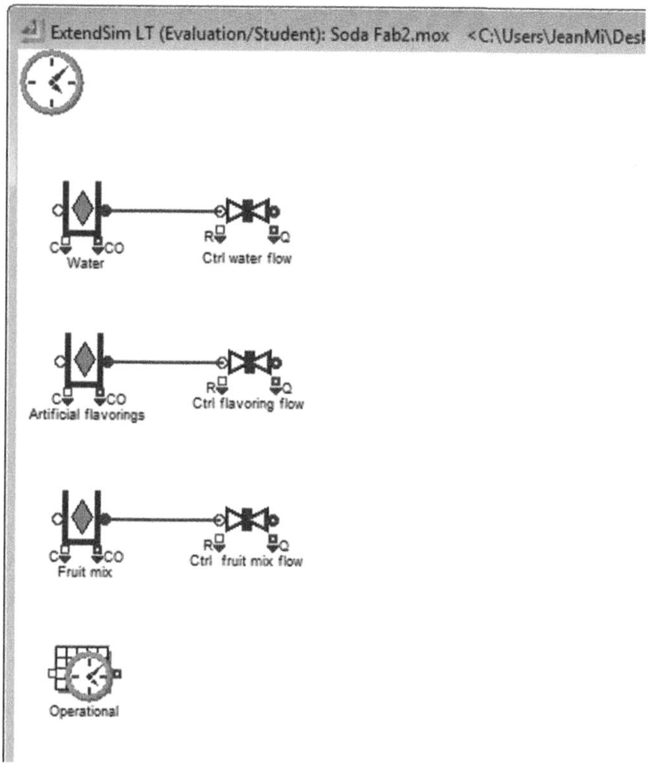

Figure 2.4. *The SHIFT, TANK and VALVE blocks in our model*

We now configure each of these blocks. For the tanks, we apply the parameters from the problem statement. Since these tanks are continually supplied by a machine which does not feature in our simulation, we can fix their initial contents and capacities as infinite (Figure 2.5).

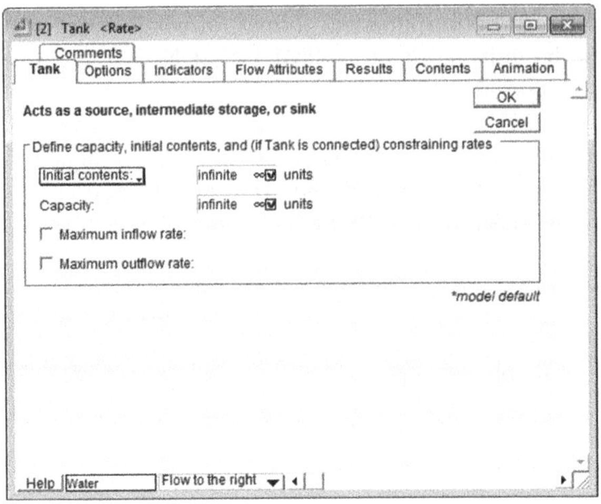

Figure 2.5. *The configuration for the 'Water' tank, with its initial contents and its capacity*

The 'Operational' schedule is assigned from the drop-down list next to USE SHIFT in the OPTIONS tab.

We equally define, using the FLOW UNIT – NEW UNIT drop-down list, the unit 'liter' for each of them.

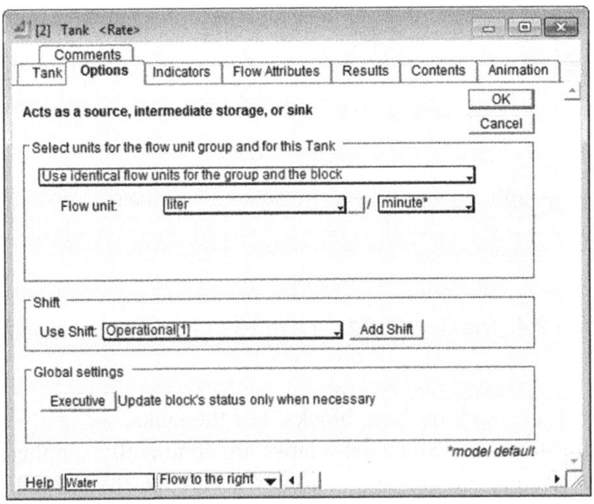

Figure 2.6. *The OPTIONS tab with the 'Operational' schedule assigned, plus the new 'liter' unit*

COMMENT 2.6.– Once a unit has been created in a block, it becomes available to all other blocks which are likely to use it.

The three VALVE blocks ('Ctrl water flow', 'Ctrl flavoring flow' and 'Ctrl fruit mix flow') will also be configured according to the figures given in the problem statement, 80 l/min for the water, 1 l/min for the artificial flavorings and 19 l/min for the fruit mix.

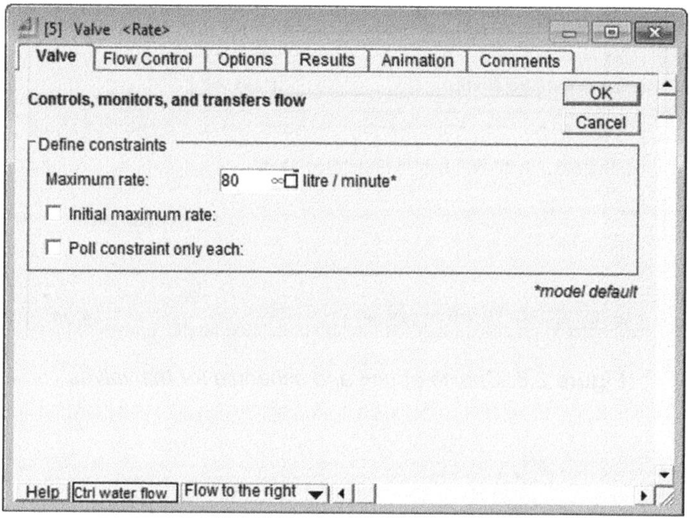

Figure 2.7. *The configuration for the valves, here the 'Ctrl water flow' valve*

COMMENT 2.7.– The names, given in inverted commas, are the labels given to the blocks in the model. They are input in the field situated at the bottom of the configuration dialog boxes, to the right of the HELP button.

We will check, under the OPTIONS tab, that the unit for each valve is indeed liter/minute and we choose the 'Operational' schedule from the drop-down list under the USE SHIFT option (Figure 2.8).

To mix the three ingredients according to the recipe (80% water, 1% flavoring and 19% fruit mix), we will use a MERGE block ('Mixing ingredients').

We start by adding an input to the block (by dragging the small black triangle on the left of the block downward) and then connecting each of the valve outputs to one of the merge inputs (Figure 2.9).

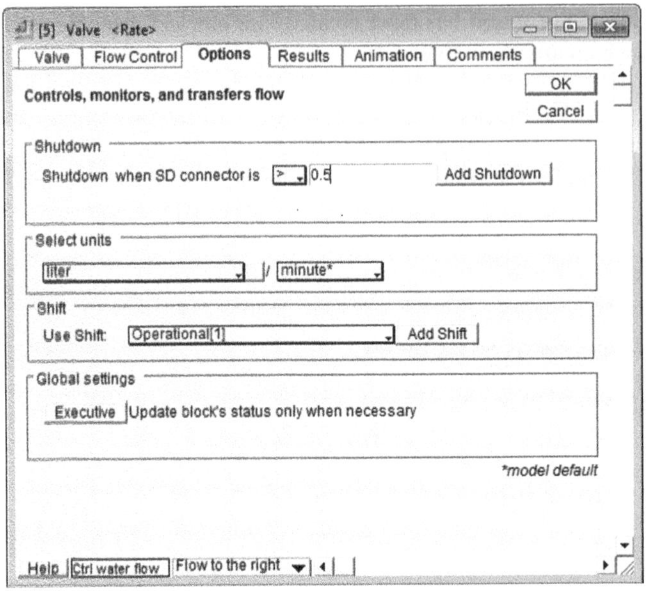

Figure 2.8. *Choice of unit and schedule for the valves*

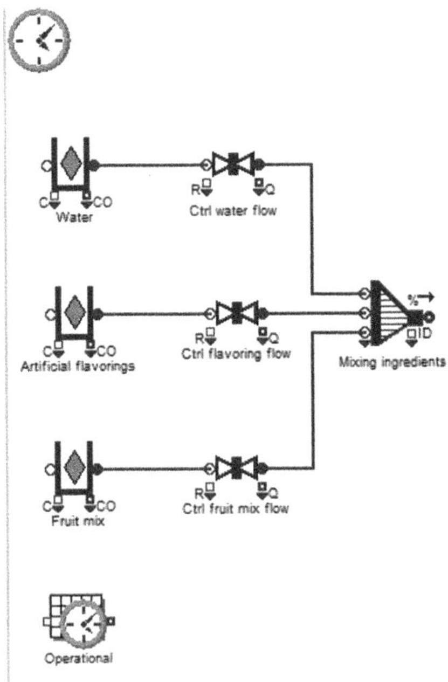

Figure 2.9. *Our model with its 3-input mixer*

The block is configured under the MERGE tab in its configuration dialog box, as shown in Figure 2.10. We select MERGE MODE: PROPORTIONAL (0.8 – 0.01 – 0.19).

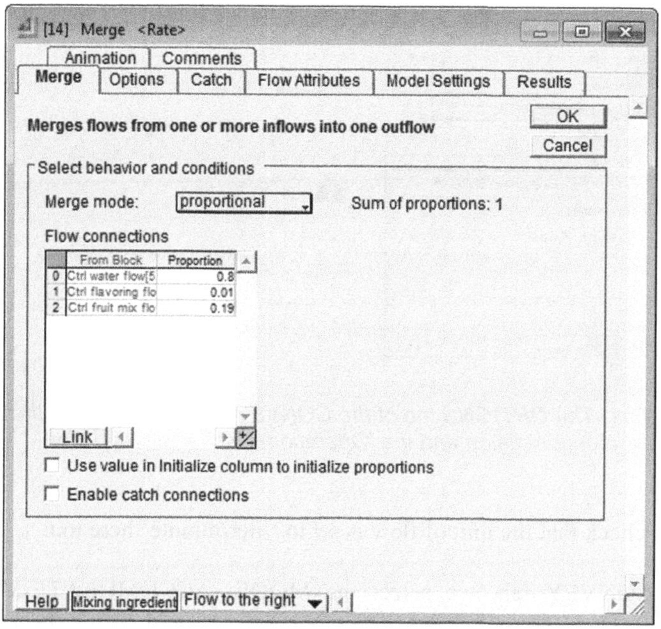

Figure 2.10. *Configuration for the MERGE block*

Under the OPTIONS tab, we should check that the units of input and output flow are 'liter/minute'.

The mixture is now sent to a heating conduit where it will be gradually heated. We will create this conduit with a CONVEY FLOW block ('Heating'), connected to the mixer's output.

The parameters for the conduit are taken from the problem statement and input under the OPTIONS and CONVEY tabs of the block.

Under OPTIONS, we create a new unit of length, 'm' for meter, using the drop-down list next to the LENGTH UNIT option. We also select the 'Operational' shift (Figure 2.11).

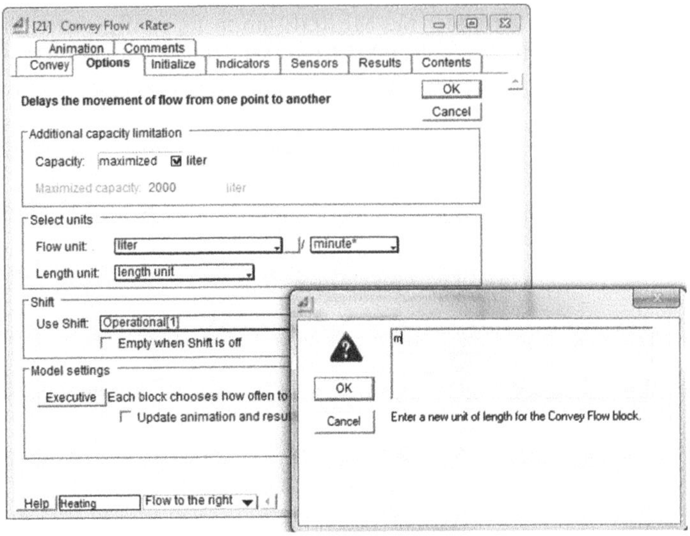

Figure 2.11. *The OPTIONS tab of the CONVEY FLOW block with the 'm' unit creation dialog and the 'Operational' schedule selection*

We also check that the unit of flow is set to 'liter/minute' here too.

In the CONVEY tab, we select the MODE: ACCUMULATE-MAXIMUM DENSITY, a TRAVEL TIME BASED ON: DELAY, and define the DELAY (10 min), the LENGTH (10 m) and the MAXIMUM DENSITY (200 l/m).

Once the mixture is at the correct temperature, we can add the sugar and send the un-bottled soda (before carbonation) on to the buffer storage tank.

The addition of sugar is based upon an identical circuit to that used for the previous components. A tank is connected to a valve which controls the flow rate of sugar, which is sent on to a mixer. The quantity of sugar is expressed in kilogram, which we convert into a volume expressed in liters to provide uniformity in the units used across the production process.

Again, all of the parameters are taken from the problem statement.

We start constructing the next part of the model by adding a TANK block ('Sugar') connected to a VALVE block ('Ctrl sugar flow') followed by a CHANGE UNITS block ('kg/liter'), itself added to a MERGE block ('Add sugar') with two inputs, as shown in Figure 2.12.

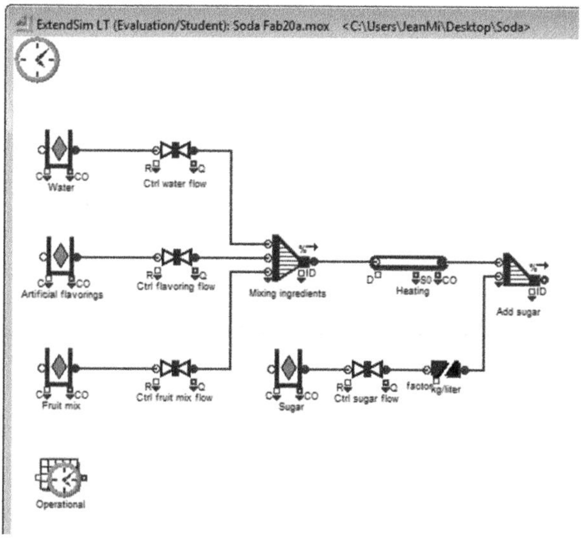

Figure 2.12. *Our model with the new TANK (Sugar), VALVE (Ctrl sugar flow), CHANGE UNITS (kg/liter) and MERGE (Add sugar) blocks*

The 'Sugar' TANK block is configured as shown in Figure 2.13 under the TANK tab. As for the three other tanks, its initial contents and capacity are both taken to be infinite.

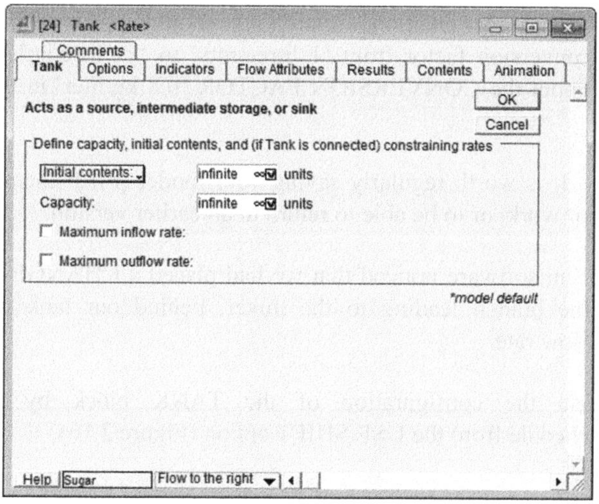

Figure 2.13. *Configuration of the sugar tank: infinite INITIAL CONTENTS and CAPACITY*

Under the OPTIONS tab, we will have to create a new unit, the 'kg', using the drop-down list for the FLOW UNIT option (Figure 2.14).

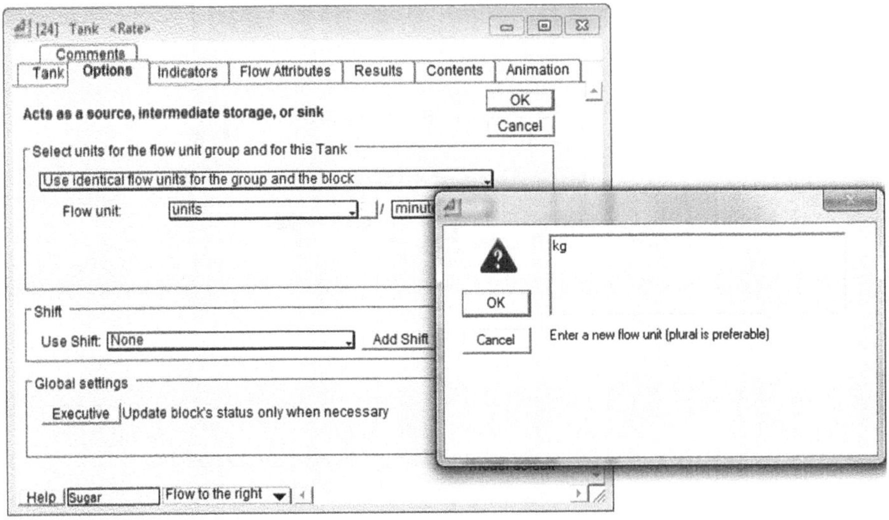

Figure 2.14. *Creating the new unit under the OPTIONS tab*

When you confirm this creation, a dialog box should appear asking if you want to change the conversion factor from '1 litre/units' to 'xx litre/kg' (Figure 2.15). Reply OK and input the CONVERSION FACTOR '0.8 kg/liter' in the new dialog box.

COMMENT 2.8.– It is worth regularly saving your model (FILE menu – SAVE) to avoid losing your work, or to be able to return to an earlier version.

The ExtendSim software noticed that we had placed a CHANGE UNITS block ('kg/liter') in the branch leading to the mixer, behind our tank and the valve controlling the flow rate.

We complete the configuration of the TANK block by selecting the 'Operational' schedule from the USE SHIFT option (Figure 2.16).

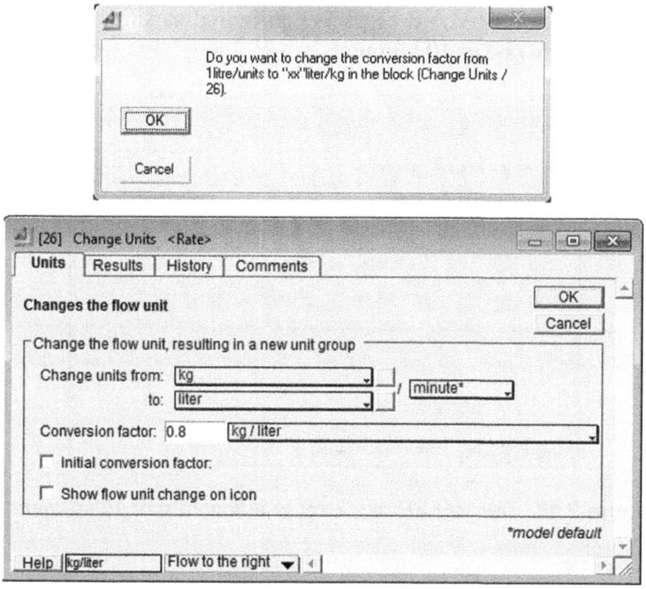

Figure 2.15. *The dialog box asking about the conversion factor (above), and the configuration window for the CHANGE UNITS block it opens (below). You can see the two units, 'kg' and 'liter', as well as the field for inputting the conversion factor: 0.8 kg/liter*

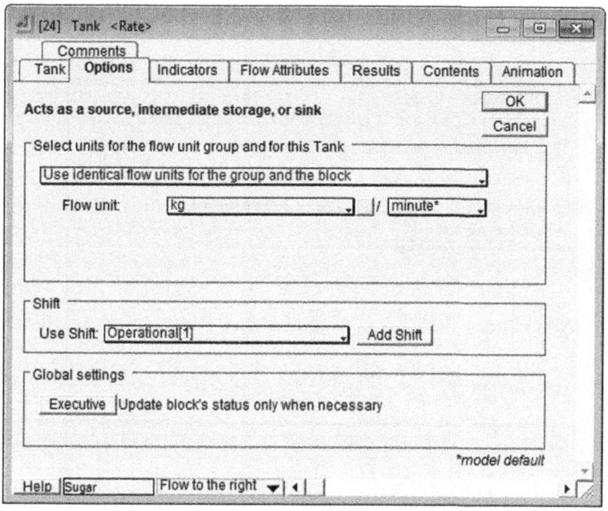

Figure 2.16. *The OPTIONS tab in the TANK block with the group unit 'kg' and the 'Operational' schedule*

The 'Ctrl sugar flow' VALVE block is configured as shown in Figure 2.17, by simply setting the flow rate to 10 kg/min.

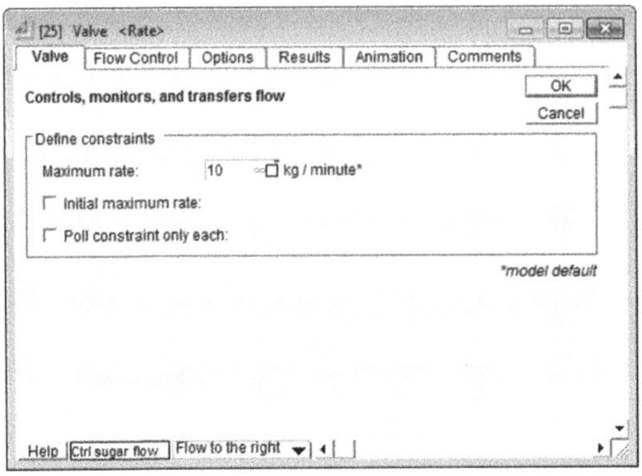

Figure 2.17. *The VALVE block set to a flow rate of 10 kg/minute*

The CHANGE UNITS block ('kg/liter') should already be configured because we have done it previously, but we can check that everything is correctly set (see Figure 2.15).

The last block in the current model adds the sugar. We will set the proportions of the two ingredients under the MERGE tab: 80% for the liquid from the heating conduit and 20% sugar (Figure 2.18), with MERGE MODE: PROPORTIONAL.

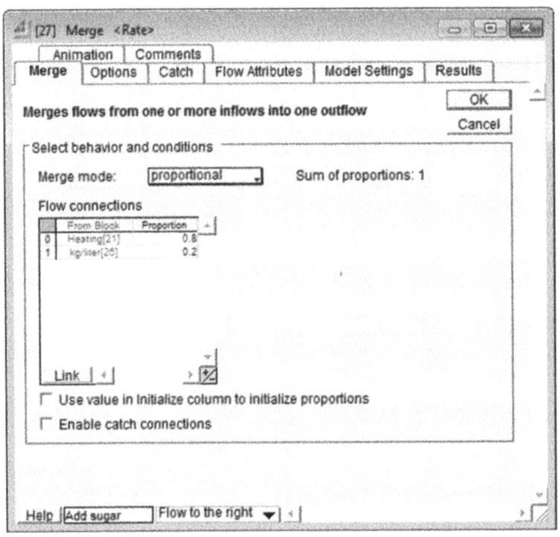

Figure 2.18. *The configuration of the 'Add sugar' MERGE block, with proportions of 0.8 and 0.2*

The mixture of un-bottled soda can now be directed toward the buffer tank that we place after the 'Add sugar' MERGE block. As specified in the problem statement, its maximum capacity is 2,000 l (Figure 2.19).

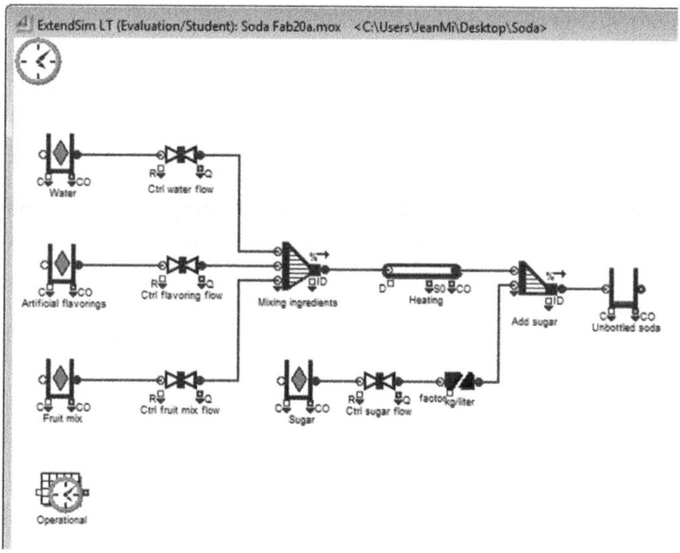

Figure 2.19. *Our model with the added un-bottled soda buffer storage TANK block*

Figure 2.20 shows the configuration of the TANK and OPTIONS tabs for the 'Un-bottled soda' storage TANK block.

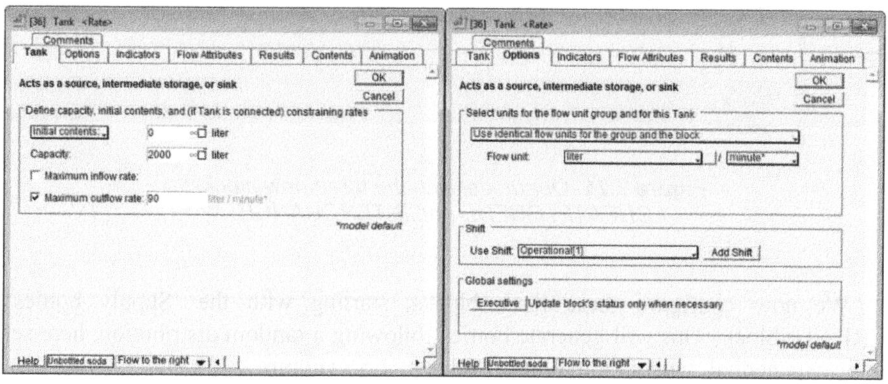

Figure 2.20. *The two dialog boxes for the TANK and OPTIONS tabs of the storage buffer tank*

The MAXIMUM OUTFLOW RATE is restricted to 90 l/min in order to meet the external constraint of one bottle of 1.5 l filled in a time that can vary between 1 and 5 s, equaling a maximum flow rate of:

$$60/1 \times 1.5 = 90 \text{ l/min.}$$

To complete this stage, we add three blocks to our model. First, we add a CREATE block ('Supply bottles') connected to a QUEUE block ('Stock bottles'), both taken from the 'Item.lix' library. These blocks deal with discrete entities.

We next connect an INTERCHANGE block ('Bottling'), found in the 'Rate.lix' library, to the un-bottled soda tank and bottle creator in order to convert the continuous liquid flow to a discrete flow of bottles.

The bottling takes place in this last block. Our model should now resemble the one in Figure 2.21.

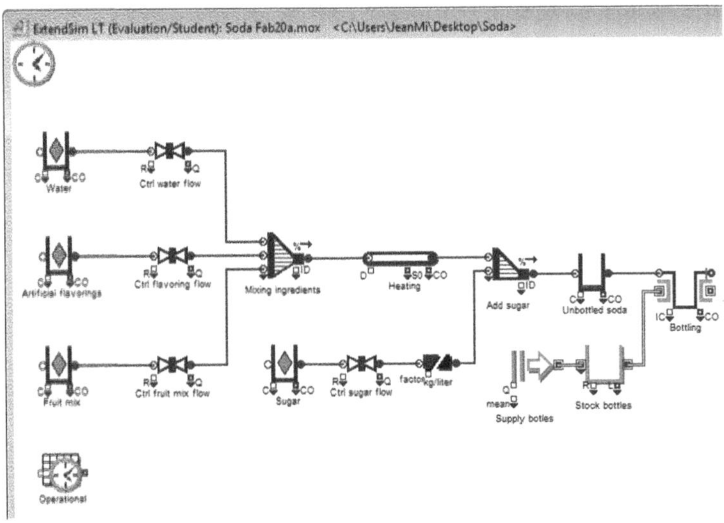

Figure 2.21. *Our model with the three new blocks: CREATE, QUEUE and INTERCHANGE*

We now configure these three blocks, starting with the 'Supply bottles' CREATE block. This will generate bottles following a random distribution, here set to be exponential, with an initial mean value of 1.5 s (Figure 2.22).

Under the ITEM ANIMATION tab, we set the 2D PICTURE to BOTTLE to stick as closely as possible to the real-life process. This symbol will spread over the links throughout the model from this point onward, unless it is later set to change (as we will see further on in the example).

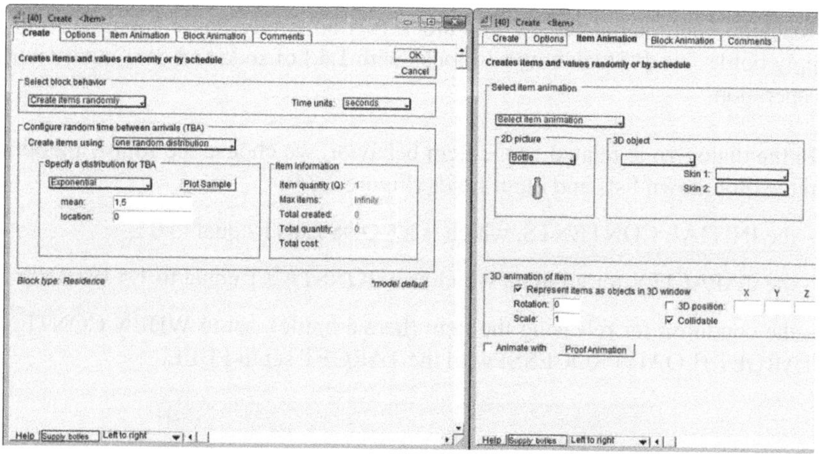

Figure 2.22. *The configuration of the 'Supply bottles' CREATE block, CREATE and ITEM ANIMATION tabs*

COMMENT 2.9.– The choice of exponential distribution should smooth the supply of bottles to the 'Bottling' INTERCHANGE block, however, it could have been replaced by a CONSTANT distribution instead.

The 'Stock bottles' QUEUE block has a 'first in, first out' (FIFO) method, and will take the role of a buffer store if the supply rate is greater than the bottling can handle.

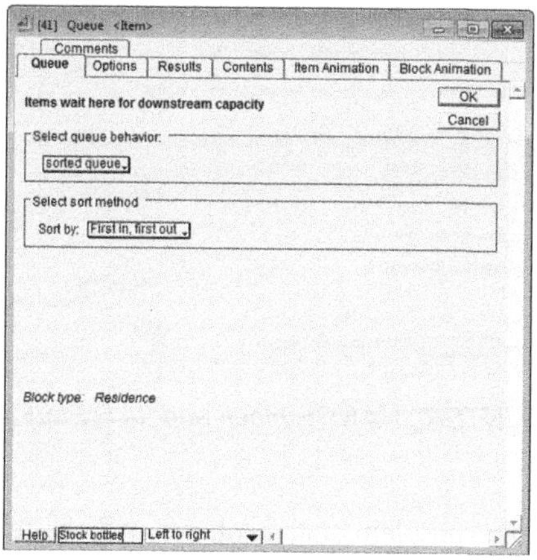

Figure 2.23. *The 'Stock bottles' QUEUE block and its parameters*

The 'Bottling' INTERCHANGE block has a slightly more complex configuration, with more parameters for us to input. Under the ITEM/FLOW tab, we define the tank relationship as ONLY EXISTS WHILE ITEM IS IN IT – essentially the machine being simulated by this block fills an empty bottle supplied from the 'Supply bottles' and 'Stock bottles' blocks with 1.5 l of soda and continually repeats this operation.

In the dialog zone related to the item behavior, we choose the following options from the drop-down lists and input fields (Figure 2.24):

– the INITIAL CONTENTS, which is a CONSTANT equal to 0 l;

– the CAPACITY for an item, which is a CONSTANT equal to 1.5 l;

– the condition for releasing the item (here a bottle), set to WHEN CONTENTS >= TARGET (LOAD PROCESS) with the TARGET set to FULL.

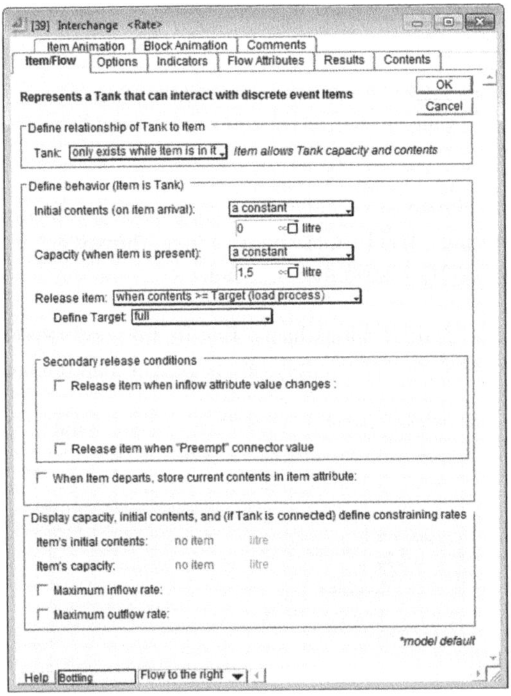

Figure 2.24. *The INTERCHANGE block and its ITEM/FLOW parameters*

If we read through the problem statement, we will see that the storage tank for un-bottled soda does not fill up instantaneously – the flow of assorted ingredients must arrive there, having left the tanks and passed through the valves, mixers and the heating conduit. We can calculate that it will take at least 20 min for the tank to start receiving the mixture. If we additionally do not want the bottling to be able to start before the tank is half-full (1,000 l), we must again add some time. We will assume that 25 min is reasonable.

COMMENT 2.10.– This value can be changed later to optimize the production process. We could also consider a trigger level dependent upon the quantity of un-bottled soda present in the tank.

It thus does not seem appropriate to start supplying bottles to the bottling machine at the same time as the un-bottled soda production line starts up. We will thus define a postponed schedule for the 'Supply bottles' creator to start, which will avoid blocking up the 'Stock bottles' QUEUE, the intermediate buffer between the CREATE block and the 'Bottling' INTERCHANGE block.

We place a supplementary SHIFT block in our model and set its parameters according to those shown in Figure 2.25.

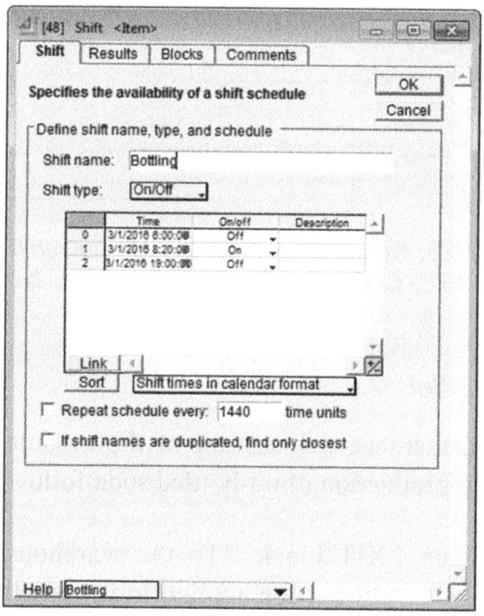

Figure 2.25. *The new SHIFT block, named 'Bottling', with its parameters: 6:00 OFF – 8:25 ON – 19:00 OFF*

We will now assign this schedule to each of the CREATE, QUEUE and INTERCHANGE blocks, under the OPTIONS tabs in their configuration dialog boxes, as shown in Figure 2.26.

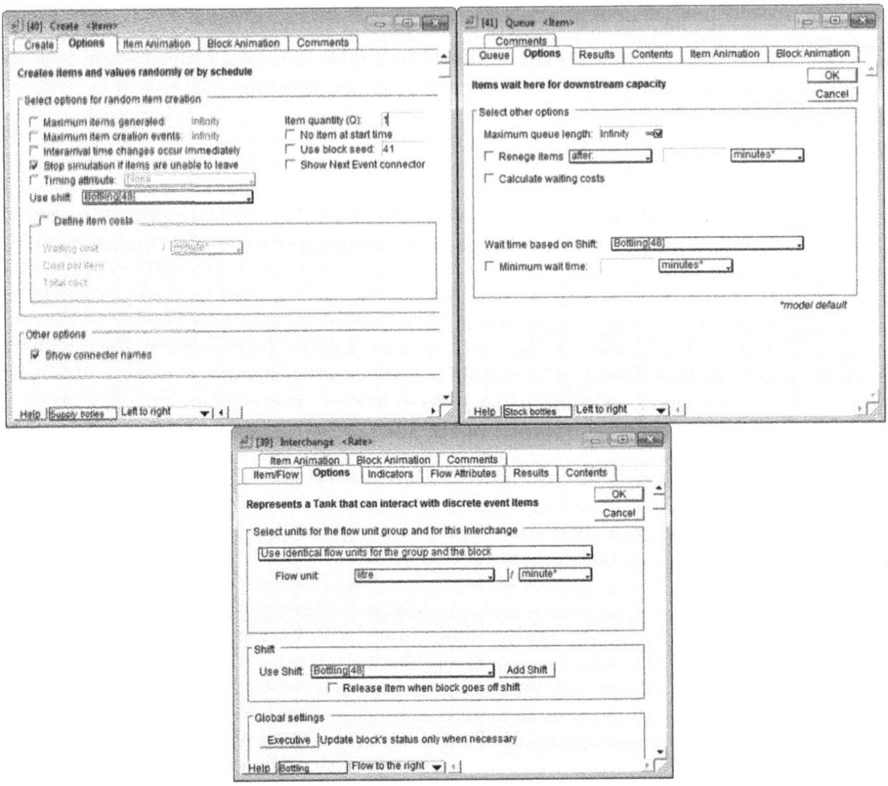

Figure 2.26. *Configuring the 'Operational' shift for the CREATE, QUEUE and INTERCHANGE blocks*

2.3.2.2. *Testing the model*

Before going any further, we will test all of the modules making up this first stage of our project – the production of un-bottled soda followed by its bottling.

To do this, we add an EXIT block ('To the warehouse') and a PLOTTER, DISCRETE EVENT block ('Production monitoring'), taken from the 'Plotter.lix' library, to monitor the operation of the production process.

Our model should now resemble the one shown in Figure 2.27.

The 'To the warehouse' EXIT block does not require any configuration, its one input is simply connected to the output from our 'Bottling' INTERCHANGE block

The PLOTTER, DISCRETE EVENT block has four inputs – we connect the first one (the top one) to the EXIT block connector labeled 'Total exited', and the second one to the 'Unbottled soda' TANK block connector labeled 'CO' (Contents).

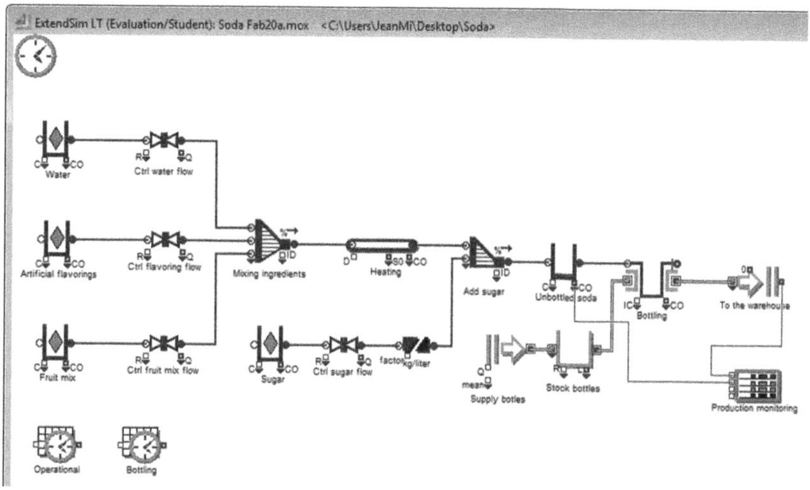

Figure 2.27. *Our model with the two new blocks: EXIT and PLOTTER, DISCRETE EVENT*

Now, we must configure the plotter. We click on the first icon from the left-hand side of the toolbar in the plotter window, TRACE PROPERTIES, and input or check the required options (Figure 2.28).

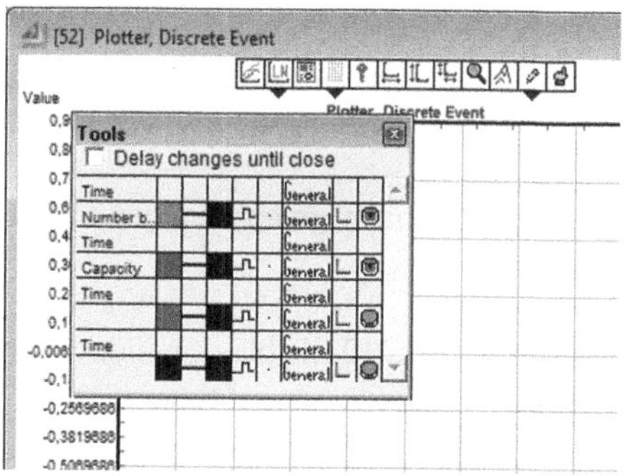

Figure 2.28. *Configuration for the plotter trace properties: 'Nb bottles'; 'Capacity (l)'; ''; ''; under TRACE NAME and SHOW TRACE unchecked for inputs 3 (green) and 4 (black). For a color version of this figure, see www.iste.co.uk/reveillac/modeling3.zip*

We next click on the third icon, OPEN DIALOG, and set the necessary parameters as in Figure 2.29.

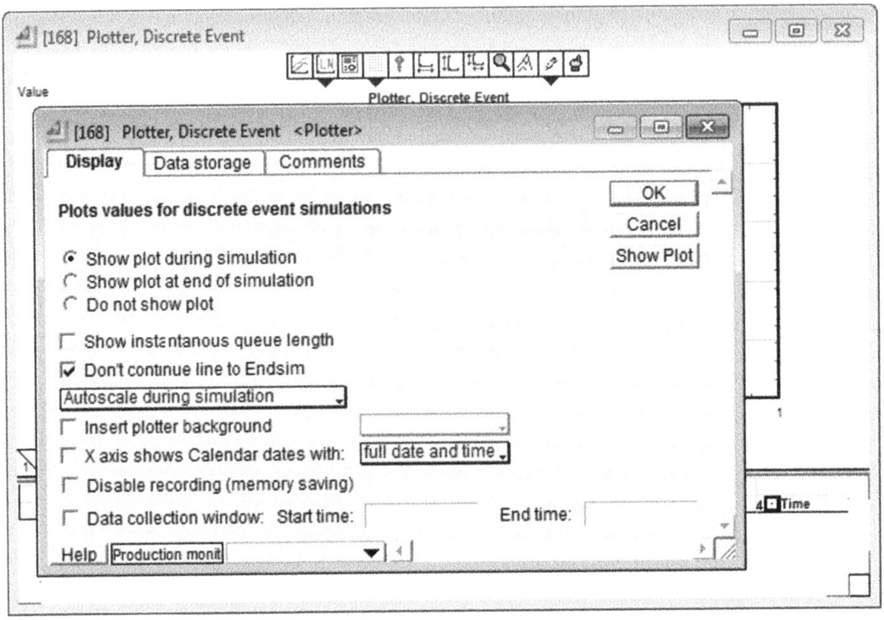

Figure 2.29. *The dialog box configuration: SHOW PLOT DURING SIMULATION, DON'T CONTINUE LINE TO ENDSIM and AUTOSCALE DURING SIMULATION*

After saving our model, we can launch the simulation, after checking 2D ANIMATION and ADD CONNECTION LINE ANIMATION in the RUN menu.

We click on the icon resembling a green triangle in the toolbar to start (see section 1.3.4).

The 'Production monitoring' PLOTTER block window should open and display the two curves monitoring the production in real time.

The number of bottles produced should verge on 26,000. This is visible on the trace and above the 'To the warehouse' EXIT block.

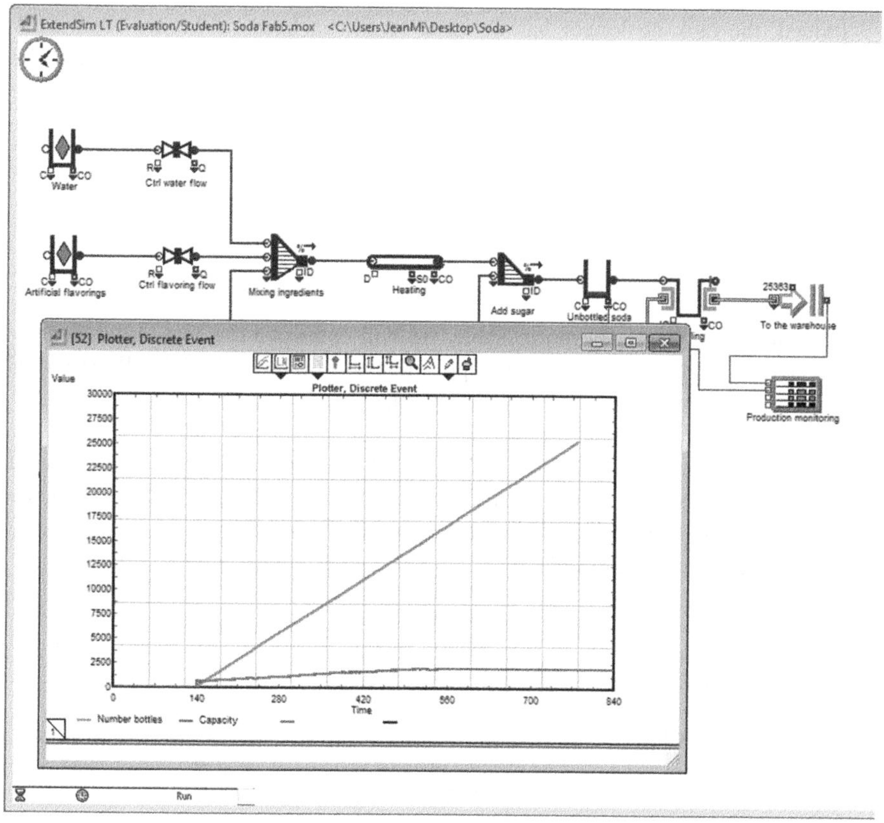

Figure 2.30. *Our model after running the simulation. We can see the plotter window with its two traces (blue: Nb bottles and red: Capacity (l) of the un-bottled soda tank). We can also make out the number of bottles produced, here 25,363, above the 'To the warehouse' EXIT block symbol. For a color version of this figure, see www.iste.co.uk/reveillac/modeling3.zip*

COMMENT 2.11.– To speed up the simulation, you can un-check 2D ANIMATION and/or ADD CONNECTION LINE ANIMATION in the RUN menu.

We could consider increasing the production by keeping the same parameters for producing un-bottled soda, but changing the speed of the bottle supply at the 'Bottling' INTERCHANGE block.

If we reduce the mean supply period from 1.5 to 1 s in the block parameters and re-run the simulation, we can see that the production yield increases to the detriment of the volume of un-bottled soda contained in the buffer tank.

Indeed, the tank no longer has the time to fill up because we are continually drawing from its contents. This situation is not actually permissible, since the bottom of the tank contains sediment from the production method which would be put into the bottles. The quality of our production would thus be worsened, generating rejects – bottles of soda unfit for consumption.

We can envisage a variety of solutions to mitigate this problem.

2.3.2.3. *Optimizing the production: solution #1*

First, to make our model more usable, we will connect a rheostat to the 'Supply bottles' CREATE block, to vary the supply rate of bottles to the INTERCHANGE block through the 'Stock bottles' QUEUE block.

In the MODEL menu, select CONTROLS and then SLIDER, a block should appear in your model. Connect its middle output to the MEAN connection of the CREATE block (Figure 2.31) and input the specified parameters (1–5 s) by clicking on the top and bottom gray fields.

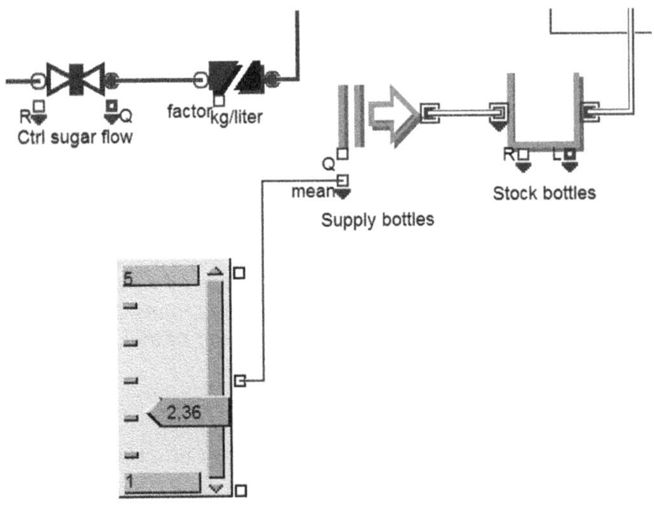

Figure 2.31. *The SLIDER block, connected and bounded (between 1 and 5), for varying the supply speed*

Run the simulation and observe the progress of the two traces monitoring production.

This approach allows us to vary the rate of supply, even during the simulation, but does not guarantee that the tank of un-bottled soda will have contents close to zero during the production.

However, if the rate is too low, for example 3.5 s, this leads to the opposite effect, with the tank of un-bottled soda exceeding its maximum capacity of 2,000 l and liable to overflow.

2.3.2.4. *Optimizing the production: solution #2*

To overcome this constraint, we will add a monitor for the contents of the un-bottled soda tank, which will fine-tune the bottling by controlling the supply rate of bottles from the machine.

We place a LOOKUP TABLE block ('Regulation') from the 'Value.lix' library in our model, whose input we connect to the 'Co' connector of the un-bottled soda tank, and whose output we connect to the 'mean' connector of our bottle creator (Figure 2.32).

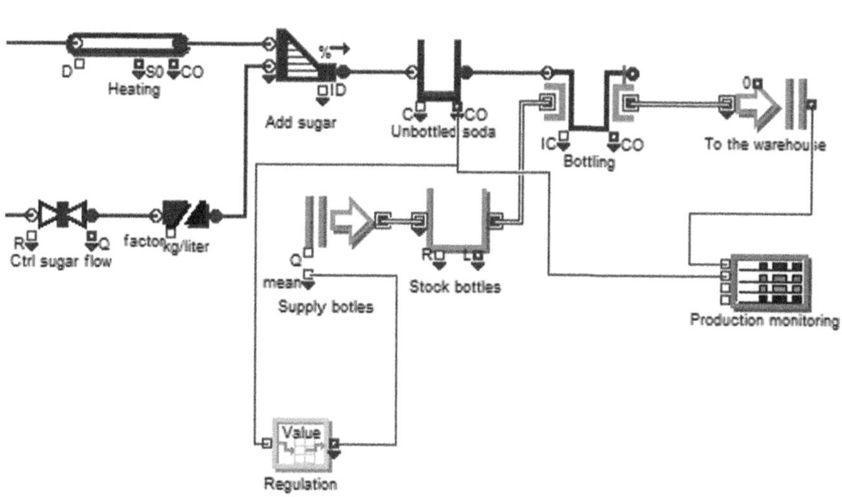

Figure 2.32. *Our model with its new LOOKUP TABLE block ('Regulation')*

We then configure the block as shown in Figure 2.33. If the contents of the tank are close to 0, we reduce the supply rate to one bottle every 5 s, however, if it approaches 2,000 we increase the rate to 1 bottle per second. The intermediate values (500, 1,000 and 1,900) provide more fine-tuning (Figure 2.33).

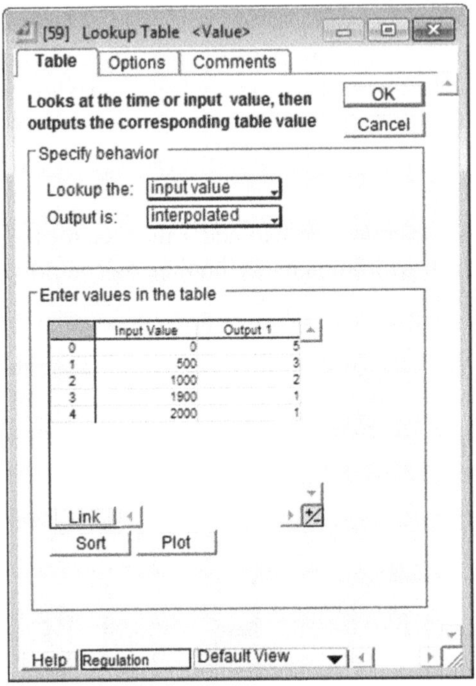

Figure 2.33. *The configuration dialog for the LOOKUP TABLE block*

By selecting INTERPOLATED from the drop-down menu for OUTPUT IS, a calculation interpolates the table's intermediate values during the simulation.

Run the simulation and observe the progress of the monitor traces.

COMMENT 2.12.–

– The boundary values of 1 and 5 bottles per second are set by the problem statement (section 2.2.1).

– The values chosen for INPUT VALUE and OUTPUT 1 in the table do not represent the only possible solution. The values displayed here have been chosen because they offer close to the optimal number of bottles produced subject to the constraints of the problem statement, while maintaining the contents of the un-bottled soda at an acceptable level (>100 and <2,000, corresponding to problems with empty and over-full tanks, respectively).

– The INPUT VALUE of 0 is required, otherwise you will receive a 'Value out of range' error message.

2.3.2.5. *Optimizing the production: solution #3*

For this solution, we will control the flow from the water tank, again by using a LOOKUP TABLE block ('Flow regulation').

We add a LOOKUP TABLE block with input connected to the 'CO' connector of the un-bottled soda tank and output connected to the 'R' connector of the valve controlling the water flow rate (Figure 2.34).

In order to improve the appearance, the block will be presented with DEFAULT VIEW REVERSED.

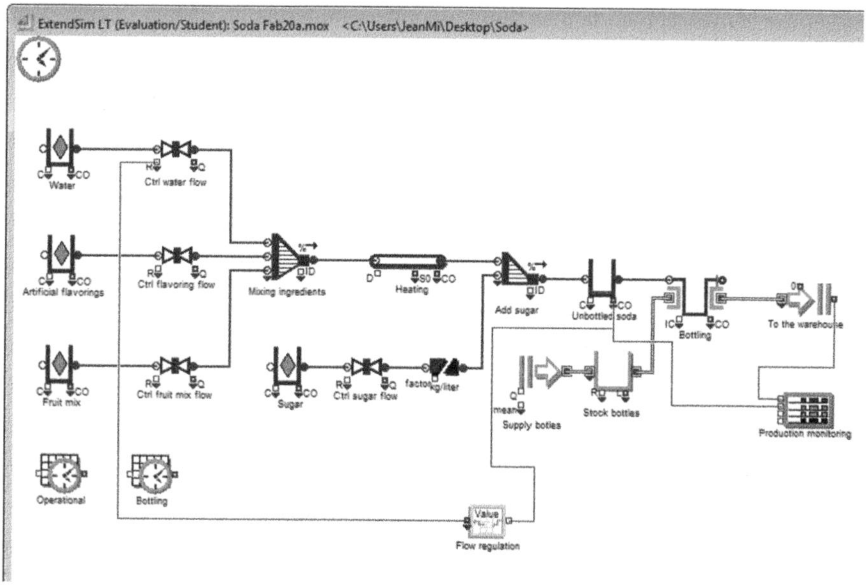

Figure 2.34. *Our model with its 'Flow regulation' LOOKUP TABLE block connected to the 'CO' and 'R' connectors*

The table configuration is given in Figure 2.35. The choice of values (0–80; 1,000–60; 2,000–0) was made with reference to the soda tank contents varying between approximately 750 and 1,500 l throughout the simulation.

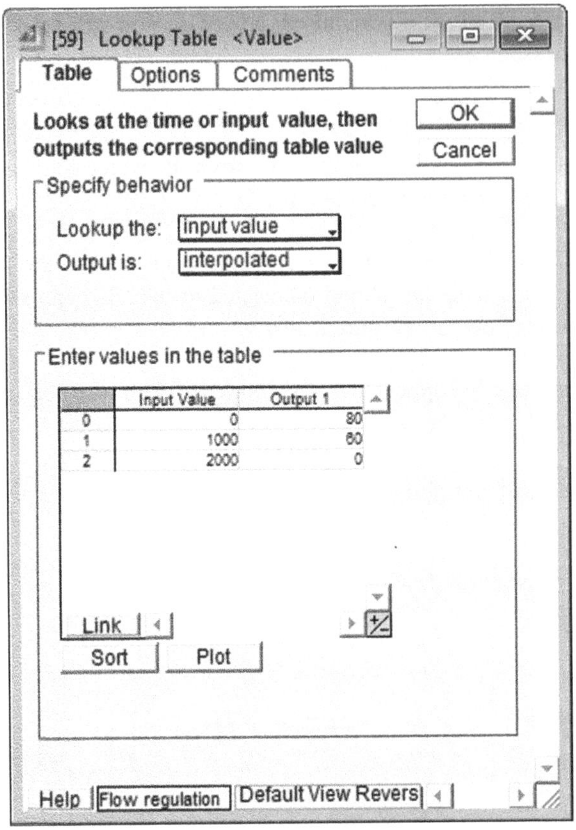

Figure 2.35. *The parameters assigned to the lookup table. Note, at bottom-right, the change in block orientation: DEFAULT VIEW REVERSED*

COMMENT 2.13.– Controlling the water flow is enough to regulate the flow from the other two tanks since the 'Mix ingredients' MERGE block is constrained to mix with defined proportions (80% water, 1% artificial flavorings and 19% fruit mix).

To improve the overview of the production monitoring, we will add a new trace to the plotter, which will show the flow from the water valve.

Open the 'Production monitoring' PLOTTER, DISCRETE EVENT block window, click on the first icon on the left in the toolbar, TRACE PROPERTIES, then input and check the necessary options, as shown in Figure 2.36.

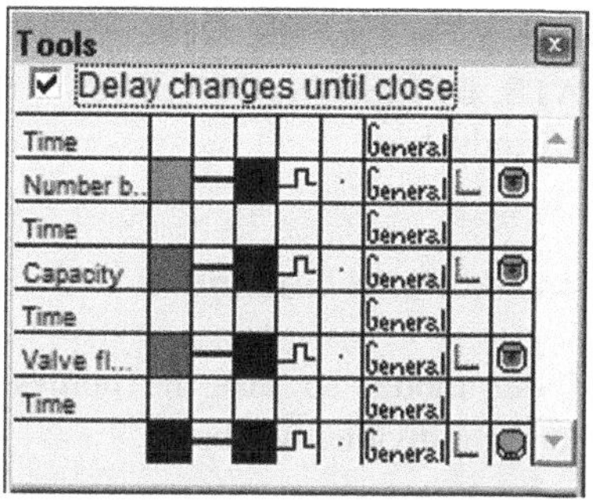

Figure 2.36. *The TRACE PROPERTIES configuration for the PLOTTER, DISCRETE EVENT block for adding the water valve flow rate ('Valve flow'). For a color version of this figure, see www.iste.co.uk/reveillac/modeling3.zip*

Run the simulation and note the considerable variation in the trace representing the tank capacity, as well as in the water flow trace (Figure 2.37).

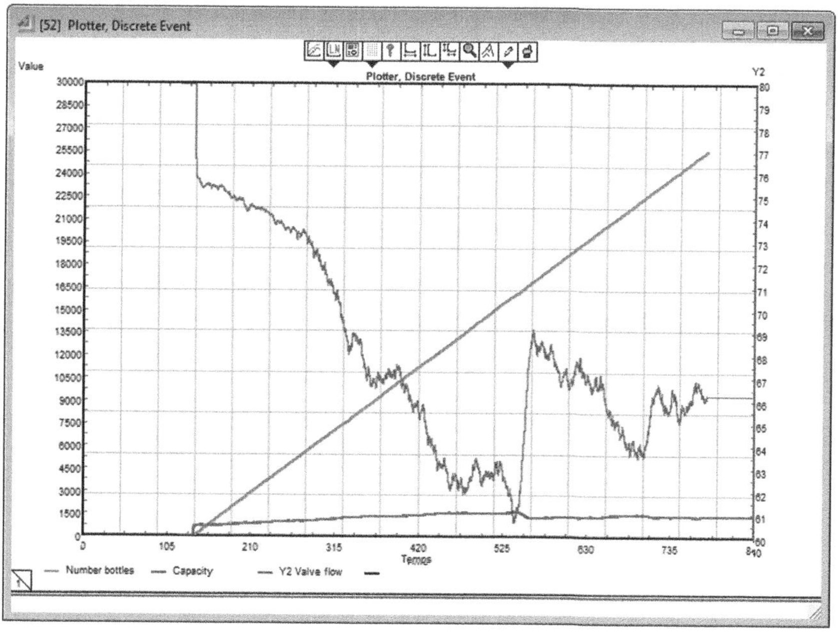

Figure 2.37. *The three traces generated by the PLOTTER, DISCRETE EVENT block in our model (blue: nb bottles; red: capacity of un-bottled soda tank in liters; and green: water flow in liters/minute). For a color version of this figure, see www.iste.co.uk/reveillac/modeling3.zip*

The three proposed solutions are not the only ones, there are numerous others using combinations of GATE, DECISION, CONSTANT, MATH and other blocks which can also manage the production.

2.3.3. *Transport, carbonation and labeling*

We will now extend our model so that the bottles are transported to the carbonation machine to then be individually labeled.

The blocks to be added are (in order, from left to right):

– CONVEY ITEM ('Convey bottles');

– QUEUE ('Stock full bottles');

– ACTIVITY ('Carbonation');

– BATCH ('Batch 1b + 1e');

– ACTIVITY ('Labeling');

– QUEUE ('Stock before packaging').

The label supply must also be added, in the form of two additional blocks: a CREATE block ('Supply labels') and a QUEUE block ('Stock labels').

All of these blocks belong to the 'Item.lix' library. Figure 2.38 shows the new blocks, placed after the bottling and all of its connections.

COMMENT 2.14.–

– The BATCH block has two inputs, one for bottles and one for labels.

– This new stage is added to the model displayed in Figure 2.21.

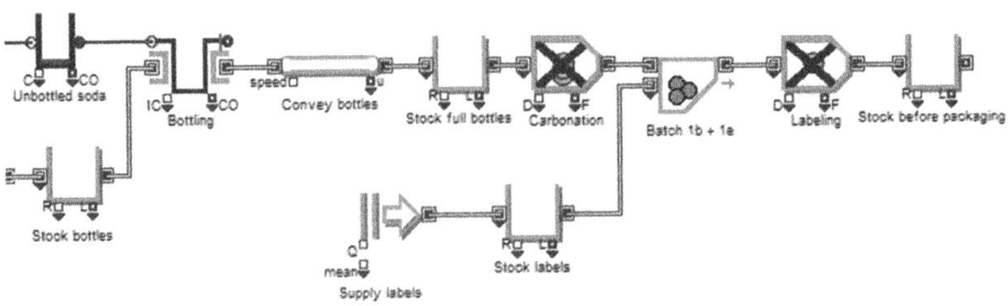

Figure 2.38. *The new blocks in the model for transport, carbonation and labeling*

Moving on to configuration, we start with the CONVEY ITEM block, which we configure according to the specifications of the problem statement.

There is no accumulation at the conveyor, it transports the bottles as they arrive. The journey lasts 20 s and the capacity of the conveyor is 100 bottles with a constant length of 0.1 m (here, we consider the diameter since the bottles are transported upright).

The shift used is identical to the previous block ('Bottling') because the rest of the model is subject to the same shift constraints (see Figure 2.25).

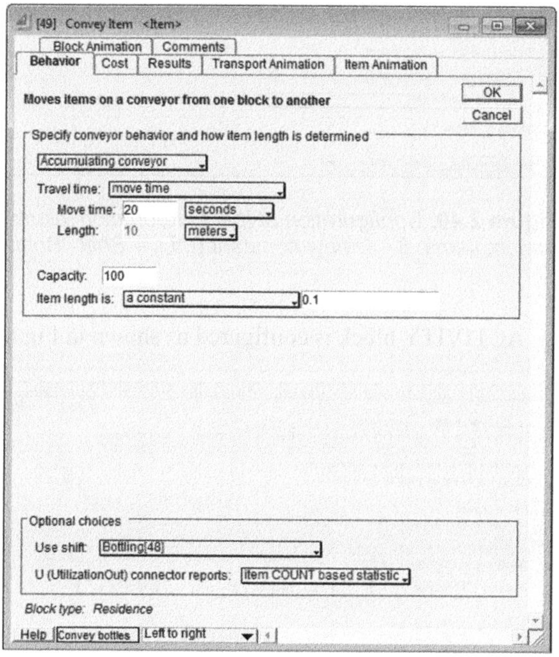

Figure 2.39. *The configuration dialog for the 'Convey bottles' CONVEY ITEM block*

The 'stock full bottles', 'stock labels' and 'stock before packaging' QUEUE blocks are of FIFO type. They are present to avoid any accumulation that could develop before the carbonation machine, after the label roller and before the packaging machines that will be added later.

The 'Carbonation' ACTIVITY block is configured as shown in 2.40. It injects the carbonic gas into the bottles and, as the configuration shows, it can process up to six bottles at a time.

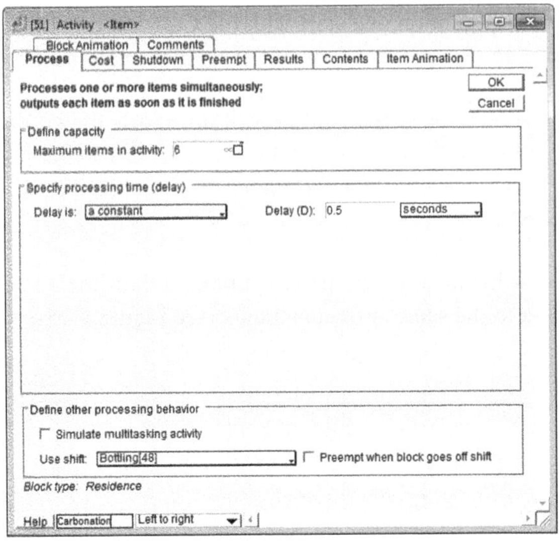

Figure 2.40. *Configuration of the 'Carbonation' block. Maximum items: 6 – Delay: constant 0.5 s – Shift: 'Bottling'*

The 'Labeling' ACTIVITY block is configured as shown in Figure 2.41.

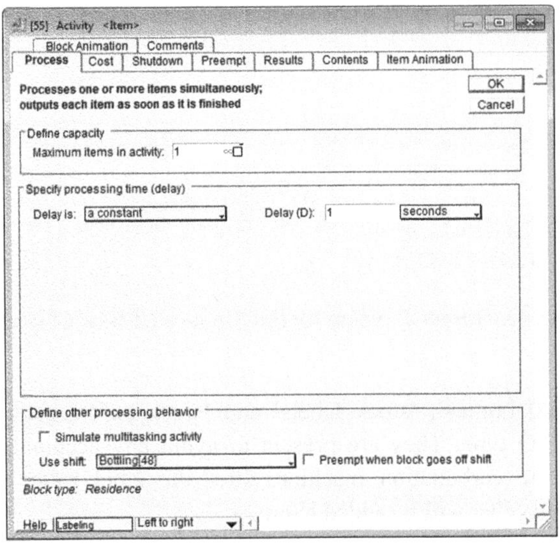

Figure 2.41. *Configuration of the 'Labeling' block. Maximum items: 1 – Delay: constant 1 s – Shift: 'Bottling'*

The role of the BATCH block is to create a new item combining one bottle with one label. We simply input the QUANTITIES, i.e. 1 for 1 (Figure 2.42).

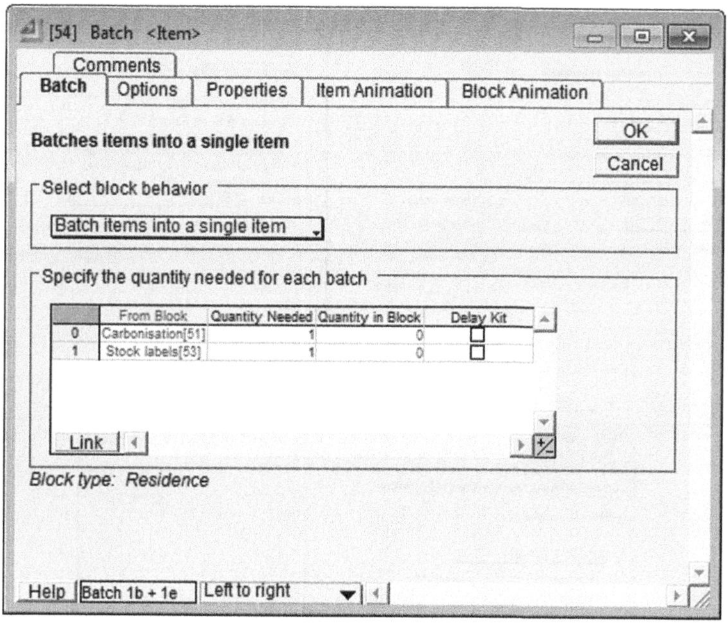

Figure 2.42. *The 'Batch 1b + 1e' BATCH block and its configuration*

We must now configure the 'Supply labels' CREATE block, which will supply one label every 1.5 s following the 'Bottling' shift, with items in the form of labels (here, we select TICKET since the item 'label' does not exist in the listed options).

Figure 2.43 shows all of the settings to input under the different tabs: CREATE, OPTIONS and ITEM ANIMATION.

2.3.3.1. *Testing the extended model*

In order to test our model, we can add an EXIT block connected to the right-hand connector of the QUEUE block containing the stock of bottles prior to packaging. Then, we simply need to run the simulation and check that it works correctly.

An hour into the production, we can see a problem with bottles backing up at the carbonation machine (the activity turns red in the SHOW 2D ANIMATION mode), which is due to the label supply rate not keeping up with the flow of bottles leaving carbonation.

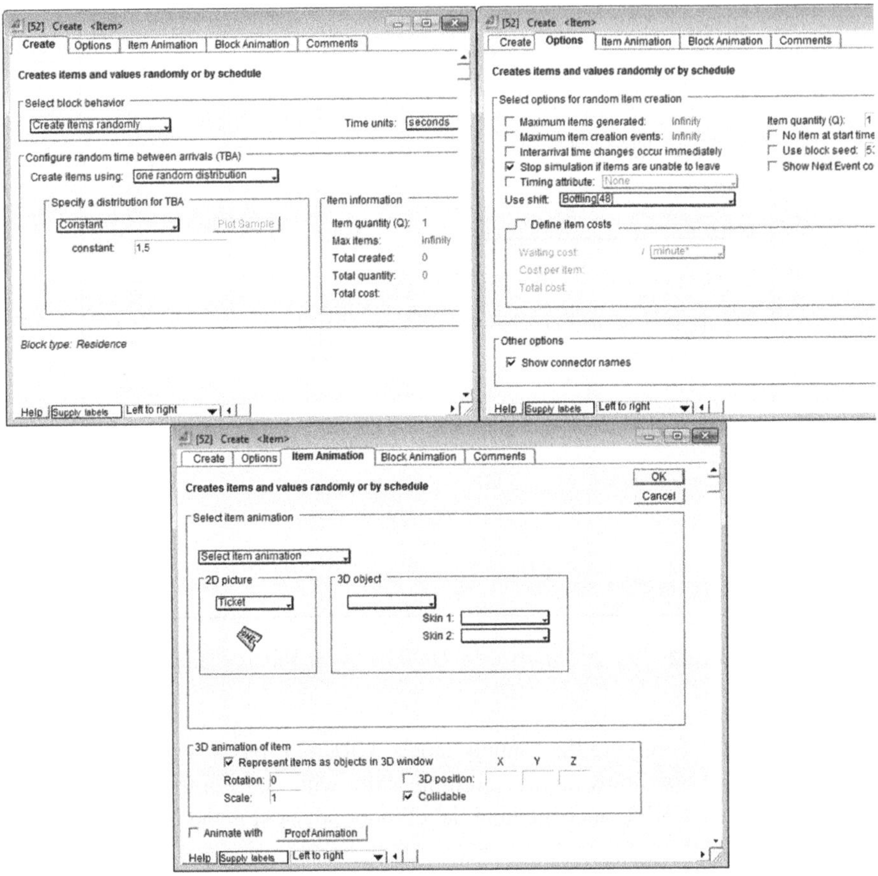

Figure 2.43. *The three tabs and the required settings for the 'Supply labels' CREATE block*

Three simple solutions present themselves:

– add a buffer store in the form of a queue between the carbonation ACTIVITY block and the BATCH block;

– marginally increase the rate of the label distribution machine by adjusting the value of the time constant on the CREATE block, for example 1.47 s (instead of 1.5 s);

– change the schedule of the label distribution machine, making it start slightly earlier in order to have a store of labels in advance, to cover possible shortcomings during production (in particular at the beginning of the day), for example starting at 8:10 am (instead of 8:20 am). Stopping the machine a little earlier would allow the buffer store of labels to be reduced at the end of the day, for example at 6:58 pm (instead of 7:00 pm).

Other possibilities exist but are more demanding, such as ensuring the label supply is monitored and regulated, or modifying the labeling system to work in batches of two or more bottles at a time.

For the remainder of our study, we go for the second of these solutions: generating the labels in a constant 1.47 s.

2.3.4. *Packaging and storage*

For these operations, our model will be split into two branches, one managing the crates and one managing the packs. Figure 2.44 shows the new extension of our model, after the bottles are stocked and ready for processing with the different packaging materials: crates, plastic film and pallets.

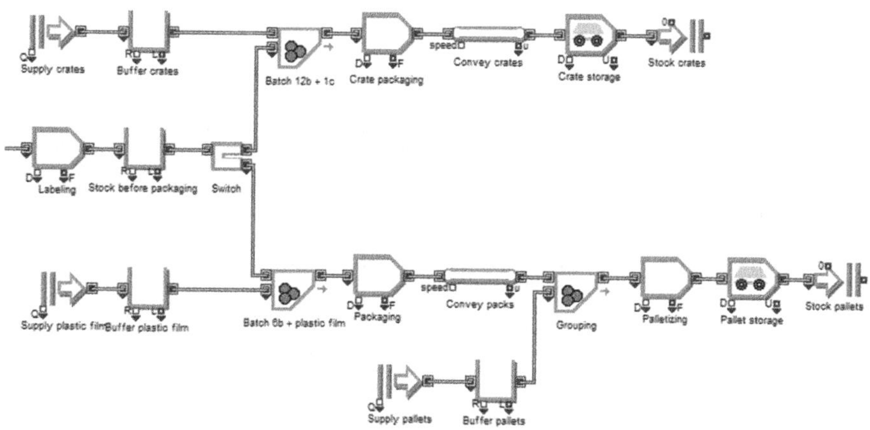

Figure 2.44. *The two branches, one for packaging and one for storage, with their supply equipment*

We construct the model using different blocks, as detailed below:

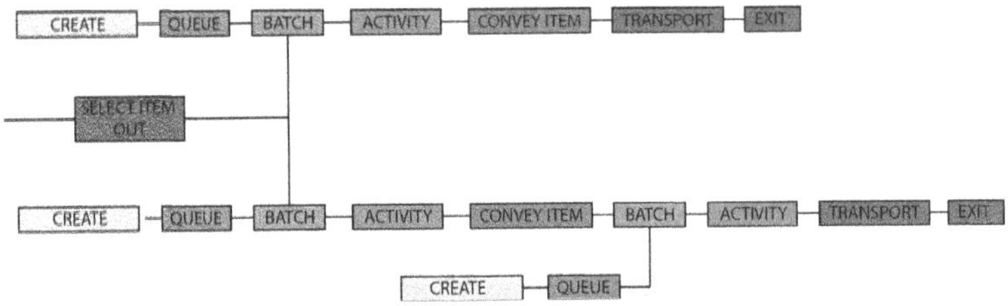

Figure 2.45. *A synoptic diagram showing all of the blocks required for extending our model. For a color version of this figure, see www.iste.co.uk/reveillac/modeling3.zip*

In order to avoid overloading this chapter, not all of the stages involved in implementing this extension will be described, except for the TRANSPORT blocks, of which there are many in this example.

The rest is summarized in the following tables, which describe all of the parameters and configurations to input into the different tabs of the dialog boxes for each of the blocks.

You can thus easily introduce and set up all of these blocks.

CREATE blocks:

Name	Tab		
Supply crates	**CREATE**	**OPTIONS**	**ITEM ANIMATION**
	Create items using: **one random distribution**	Use shift: **Bottling**	Select item animation: **Box_open**
	Specify a distribution for TBA: **constant -150 seconds** *(to be adjusted)**		
Supply plastic film	**CREATE**	**OPTIONS**	**ITEM ANIMATION**
	Create items using: **one random distribution**	Use shift: **Bottling**	Select item animation: **Paper**
	Specify a distribution for TBA: **constant -10 seconds** *(to be adjusted)**		
Supply pallets	**CREATE**	**OPTIONS**	**ITEM ANIMATION**
	Create items using: **one random distribution**	Use shift: **Bottling**	Select item animation: **Pallet_empty**
	Specify a distribution for TBA: **constant - 15 minutes** *(to be adjusted)**		

* *After running the simulation and analyzing the results, it is possible that the flow rate of each creator will need to be adjusted in order to produce the largest possible number of crates and pallets, without blockage or bursting, across the whole production line.*

QUEUE blocks:

Name	Tab
Buffer crates	QUEUE
	Sort by: First in, first out
Buffer plastic film	QUEUE
	Sort by: First in, first out
Buffer pallets	QUEUE
	Sort by: First in, first out

BATCH blocks:

Name	Tab
Batch 12b + 1c	BATCH
	Quantity needed from 'Buffer crates' block: 1
	Quantity needed from switch: 12
Batch 6b + plastic film	BATCH
	Quantity needed from switch: 6
	Quantity needed from 'Buffer film' block: 1
Grouping	BATCH
	Quantity needed from 'Convey pack' block: 72
	Quantity needed from 'Buffer pallets' block: 1

ACTIVITY blocks:

Name	Tab
Crate packaging	PROCESS
	Delay is: a constant – Delay (D): 6 seconds
	Maximum items in activity: 1
	Use shift: Bottling
Packaging	PROCESS
	Delay is: a constant – Delay (D): 1.2 seconds
	Maximum items in activity: 1
	Use shift: Bottling
Palletizing	PROCESS
	Delay is: a constant – Delay (D): 20 seconds
	Maximum items in activity: 1
	Use shift: Bottling

CONVEY ITEM blocks:

Name	Tab
Convey crates	BEHAVIOR
	Non-accumulating conveyor
	Travel time: speed and distance
	Move time: 6 seconds
	Length: 50 meters
	Speed: 5 meters/second
	Capacity: 64
	Item length: based of length & capacity
	Use shift: Bottling
Convey packs	BEHAVIOR
	Non-accumulating conveyor
	Travel time: speed and distance
	Move time: 5 seconds
	Length: 20 meters
	Speed: 4 meters/second
	Capacity: 40
	Item length: based of length & capacity
	Use shift: Bottling

SELECT ITEM OUT block:

Name	Tab
Switch	OPTIONS
	To block 'Batch 12b + 1c' - Probability: 0.15
	To block 'Batch 6b + plastic film' - Probability: 0.85

Table 2.1. *The parameters and attributes to input for the new blocks in our model*

For the two TRANSPORT blocks (used when items are moved from one point to another, whether defined by a calculated speed or distance), we input the following parameters, as shown in Figure 2.46:

– CAPACITY: INFINITY

– USE SHIFT: 'Bottling'

– TRAVEL TIME: MOVE TIME

– Move time:

- 30 SECONDS for the 'Crate storage' block.

- 4 MINUTES for the 'Pallet storage' block.

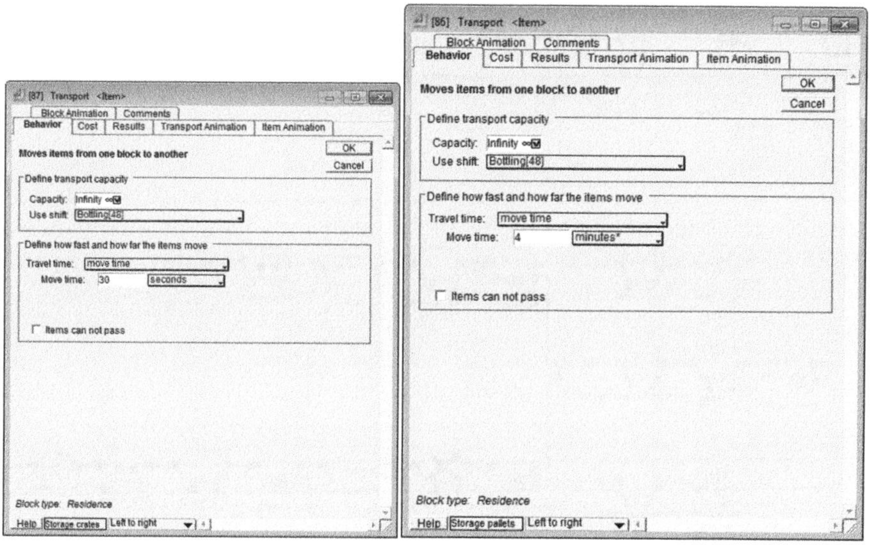

Figure 2.46. *The different parameters to be input under the BEHAVIOR tabs of the 'Crate storage' and 'Pallet storage' blocks*

COMMENT 2.15.– The two EXIT blocks have no specific parameters, their inputs are simply connected to the outputs of each of the TRANSPORT blocks.

Once all of the blocks are in place, we should have a model similar to the one shown in Figure 2.47.

2.3.4.1. *Testing and analysis*

We now run our simulation and analyze the results.

With the parameters indicated previously and shown in Table 2.1, our production line should, by the end of a day, have produced around 250 crates and 40 pallets.

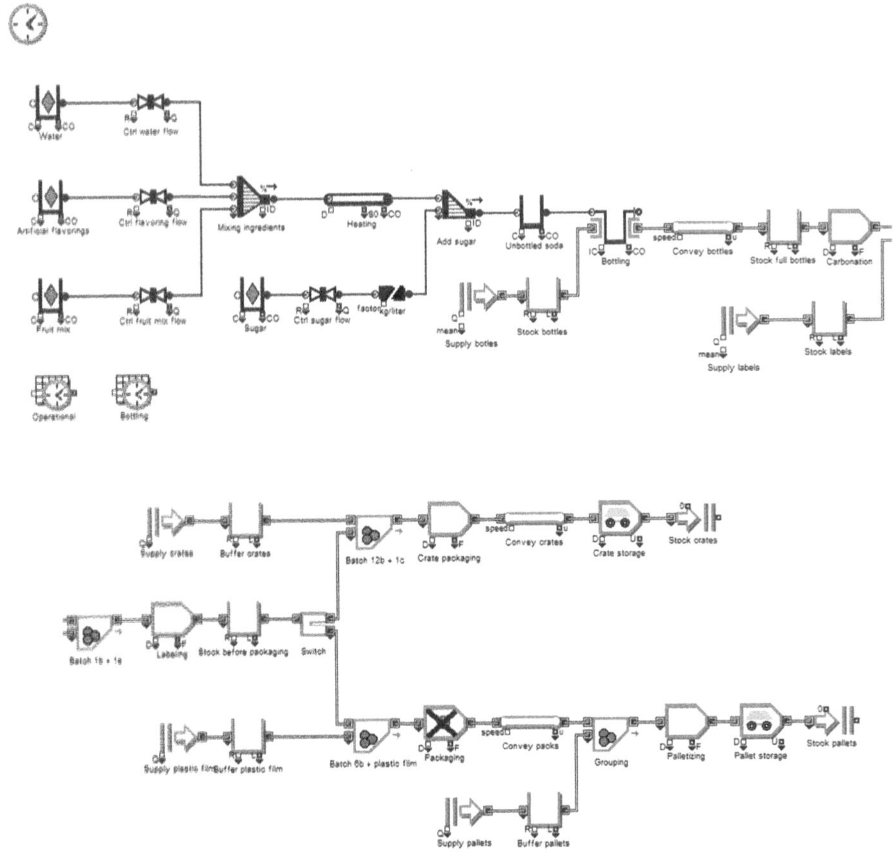

Figure 2.47. *Our complete model. In order to fit on one page, this figure shows the model split into two parts, which are actually connected and adjacent to one another (The 'Carbonation' ACTIVITY and 'Stock labels' QUEUE blocks are connected to the 'Batch 1b + 1e' BATCH block.)*

We focus our analysis on the following parts:

– the buffer stores for each of the QUEUE blocks of our three CREATE blocks, 'Supply crates', 'Supply plastic film' and 'Supply pallets';

– the 'Stock before packaging' buffer store.

We can see a severe imbalance, in particular in the stock of bottles ready for packaging (of several thousand bottles).

This suggests that the packaging and storage parts of our production line are slow and cannot keep up with the rate set by the rest of the process.

We are unable to fine-tune the ACTIVITY blocks, which must meet the external settings, however, we are able to vary the supply rate from our CREATE blocks.

COMMENT 2.16.–

– For more precise analysis of the results, you can place one or more discrete plotters connected to certain strategic points, such as the ‘L’ connectors of the QUEUE blocks or the ‘Total exited’ connectors on the EXIT blocks. We can also add a STATISTICS block (‘Value.lix’ library which will monitor the process and provide multiple figures of interest. All of this can be seen in Figure 2.48.

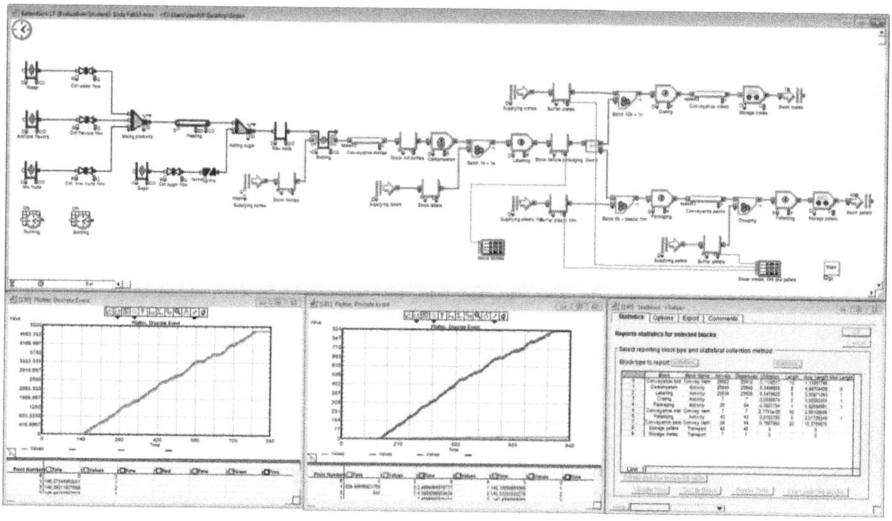

Figure 2.48. *The model with two PLOTTER, DISCRETE EVENT blocks monitoring the buffers and a STATISTICS block (at bottom-right). For a color version of this figure, see www.iste.co.uk/reveillac/modeling3.zip*

– You can also post the dialog results from each of the buffer stores, which will show you the development of each of the parameters during the simulation (Figure 2.49).

– It is also possible to extract these elements from their dialog box, posting them in the model, as presented in section 1.3.10 of this book (Figure 2.50).

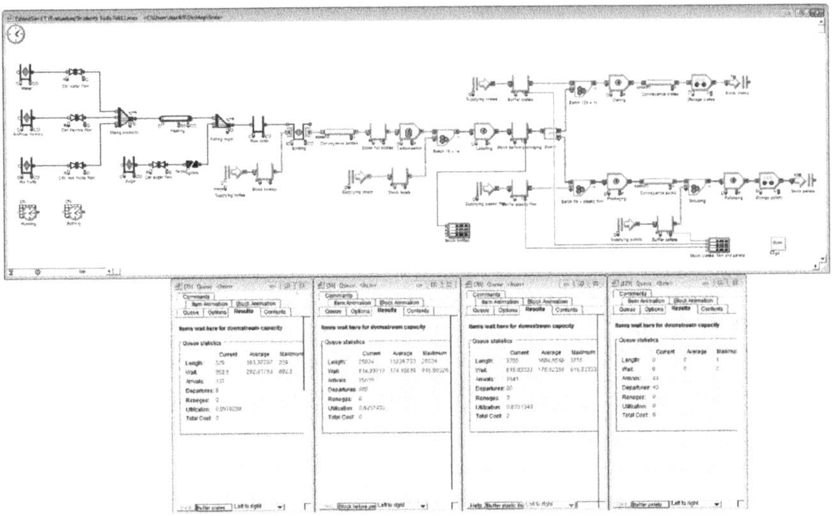

Figure 2.49. *The model with the dialogs from the RESULTS tabs for each of the QUEUE blocks from the packaging and storage part of our model*

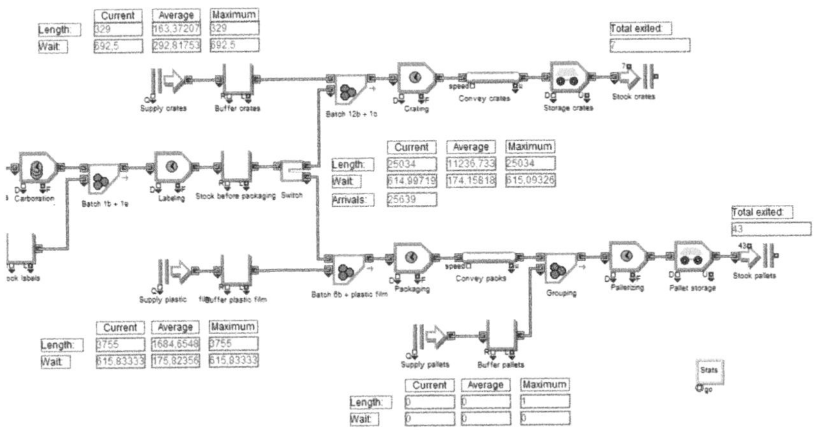

Figure 2.50. *All of the cloned parameters for the QUEUE and EXIT blocks placed in the model*

As described above, we input the following values into the three CREATE blocks:

– 'Supply crates' block: CONSTANT - 112 SECONDS;

– 'Supply plastic film' block: CONSTANT – 10.55 SECONDS;

– 'Supply pallets' block: CONSTANT – 12.5 MINUTES.

Re-running the simulation, we can see a clear change, with more balanced stocks and 331 crates and 50 pallets produced. The production has thus been improved.

COMMENT 2.17.– In reality, it would certainly be possible to fine-tune the conveyors, the various packaging robots, or even the travel times of the transport operations. This would impact upon the production and the results might be improved yet further.

2.3.5. *Maintenance and cleaning*

At this stage, we will add the daily maintenance and cleaning (washing and rinsing) routines.

If we look back at the problem statement, we can see that our production line undergoes a maintenance routine every morning and a full clean (washing – rinsing) at the end of every day.

To simulate these two routines, we will create a process parallel to our production line in which we will define two ACTIVITY blocks, 'Maintenance' and 'Cleaning'.

These will be linked to two SHIFT blocks, respectively:

– 'Maintenance schedule' from 6:00 to 8:00 am;

– 'Cleaning schedule' from 7:00 to 8:00 pm.

The resources – four technicians and five specialist workers – will be placed into two groups, 'Technicians' and 'Specialist workers', made using RESOURCE POOL blocks.

Each of these two groups has its own SHIFT block, following the work times detailed in the problem statement:

– Technician schedule: 6:00 am–1:00 pm;

– Specialist worker schedule: 1:00 pm–8:00 pm.

A CREATE block will generate an event once a day, which will trigger first the maintenance and then the cleaning.

Figure 2.51 shows all of the blocks joined together. Two RESOURCE POOL RELEASE blocks are present to allow the technicians and workers to be freed from their work on this production line to attend to other tasks during the rest of the workday (after 8:00 am for the technicians and before 7:00 pm for the workers).

The QUEUE blocks allocate the resources to the maintenance and cleaning activities.

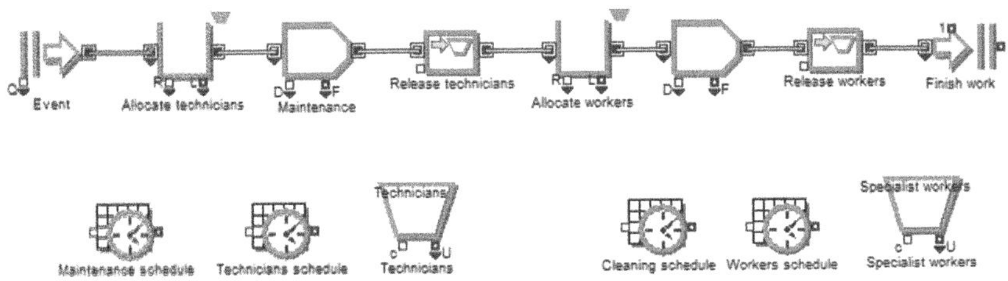

Figure 2.51. *All of the blocks to be placed in our model to ensure the daily maintenance and cleaning of the production line*

The configuration of the 'Event' CREATE block is completed in the CREATE, OPTIONS and ITEM ANIMATION tabs.

A CONSTANT of 840 MINUTES (one factory workday) is assigned as the distribution parameter. This ensures the generation of one event per day, further limited by the MAXIMUM ITEMS GENERATED, here 1.

The calendar does not need to be specified, by default it will follow the parameters of the simulation, which takes place over 1 day (Figure 2.52).

COMMENT 2.18.– A factory calendar could be defined using a SHIFT block, nothing in this example prevents the 'Event' CREATE block from using it. This could be useful if the duration of the simulation is modified and it runs over several days.

In order to be more realistic, it is worth going into the ITEM ANIMATION tab and selecting TOOL (wrench) from SELECT ITEM ANIMATION.

We must now define the four schedules we need in the RESOURCE POOL and ACTIVITY blocks (Figure 2.53).

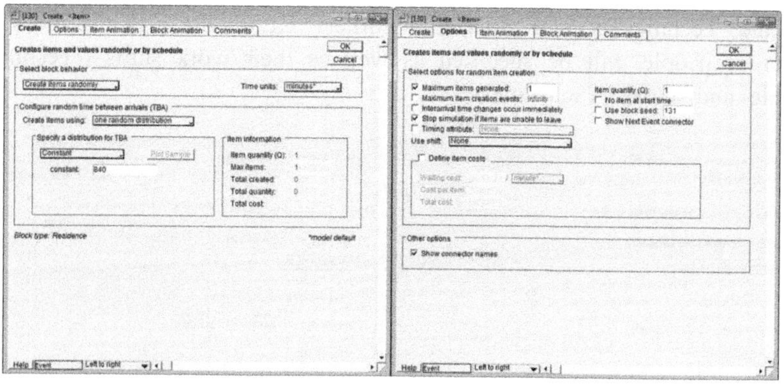

Figure 2.52. *The parameters in the CREATE and OPTIONS tabs of the 'Event' CREATE block*

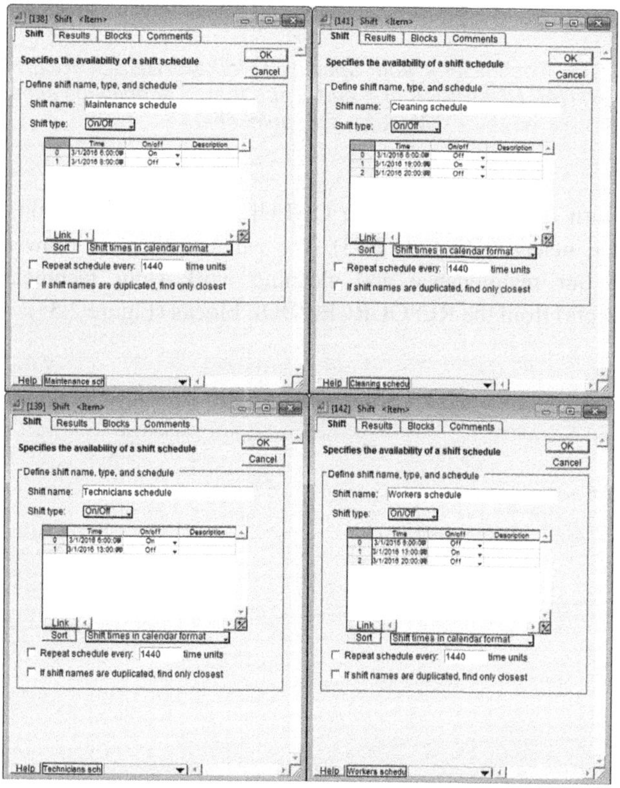

Figure 2.53. *The four SHIFT blocks: 'Maintenance schedule', 'Cleaning schedule', 'Technicians schedule' and 'Workers schedule'*

In the ‘Technicians’ and ‘Specialist workers’ RESOURCE POOL blocks, the number of people will be specified as well as their work shifts (‘Technicians schedule’ and ‘Workers schedule’).

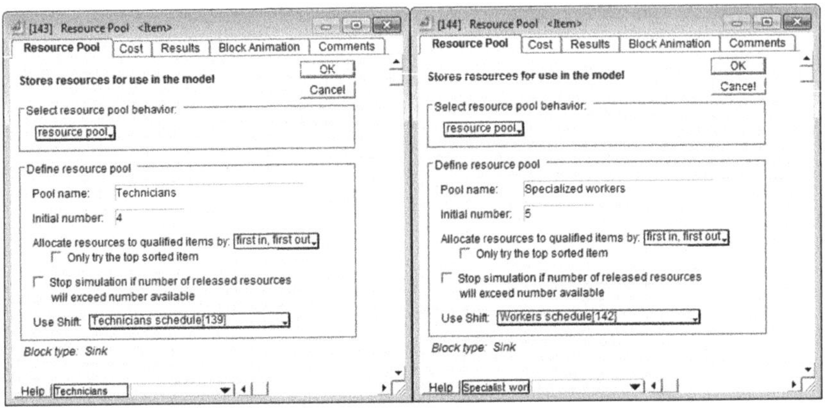

Figure 2.54. *The ‘Technicians’ and ‘Specialist workers’ RESOURCE POOL blocks with their INITIAL NUMBERS of 4 and 5 and their ‘Technicians schedule’ and ‘Workers schedule’ shifts*

We now turn our attention to the QUEUE blocks that take on RESOURCE POOL QUEUE behavior (Figure 2.55). We will use them to draw the necessary resources for our maintenance and cleaning work (four technicians and five specialist workers) from the RESOURCE POOL blocks (Figure 2.55).

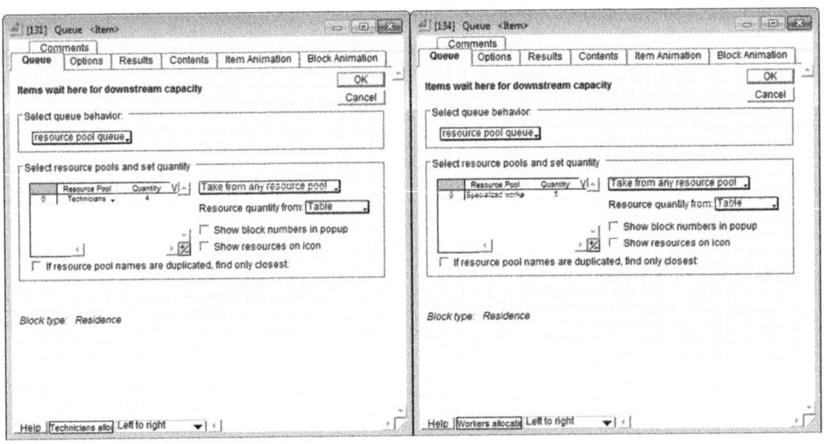

Figure 2.55. *The two QUEUE blocks, ‘Allocate technicians’ and ‘Allocate workers’, with their RESOURCE POOL and QUANTITY parameters*

The ACTIVITY blocks represent the maintenance and cleaning duties, which last 120 and 60 min, respectively, and are dependent upon the 'Maintenance schedule' and 'Cleaning schedule' shifts (Figure 2.56).

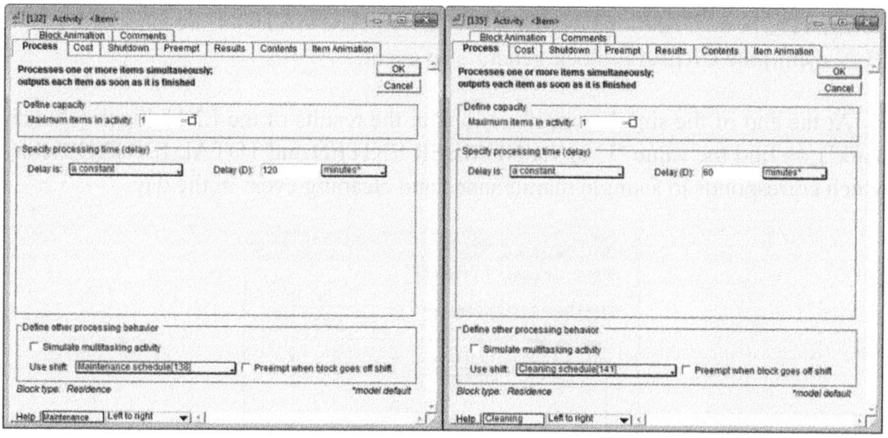

Figure 2.56. *The two ACTIVITY blocks for the maintenance and cleaning operations*

The four technicians and five workers must be freed when they are not dealing with the preventative maintenance or the cleaning, which leads us to add two RESOURCE POOL RELEASE blocks.

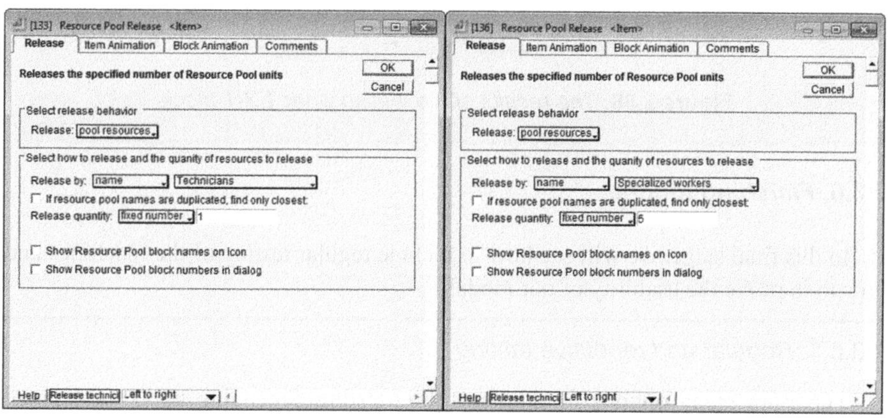

Figure 2.57. *The two RESOURCE POOL RELEASE blocks ('Release technicians' and 'Release workers')*

As for the EXIT block ('Finish work'), no configuration is required.

2.3.5.1. *Running and testing*

We can now run the model. At the start of the simulation, we can see (if 2D ANIMATION and ADD CONNECTION LINE ANIMATION are checked in the RUN menu) the CREATE block generate an event.

At the end of the simulation, if we look at the results of the EXIT block ('Finish work') we find the value '1' in the NUMBER EXITED and TOTAL EXITED fields, which corresponds to a single maintenance and cleaning event in the day.

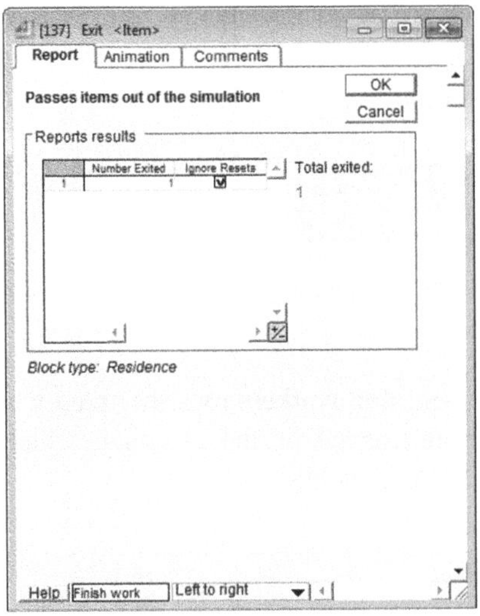

Figure 2.58. *The results of the 'Finish work' EXIT block*

2.3.6. *Finishing touches*

In this final stage, we will see how to include regular testing of the machines and how to improve the usability of our model.

2.3.6.1. *Regular and repetitive testing*

There are many options for implementing testing depending upon the type of block we are working with. Three examples applicable to our model will be presented here:

– testing the 'Ctrl water flow' valve;

– testing and fine-tuning the ‘Crate packaging’ robot;

– testing the ‘Palletizing’ robot.

In all three cases, the SHUTDOWN block from the Item.lix’ library will be used.

COMMENT 2.19.– Although this block is called ‘SHUTDOWN’, it is usable in other situations such as repetitive testing, maintenance, programmed stops, etc.

2.3.6.1.1. Water flow rate valve

We introduce a SHUTDOWN block (‘Maintenance valve’) and connect its output to the ‘R’ connector of the ‘Ctrl water flow’ VALVE block, as shown in Figure 2.59.

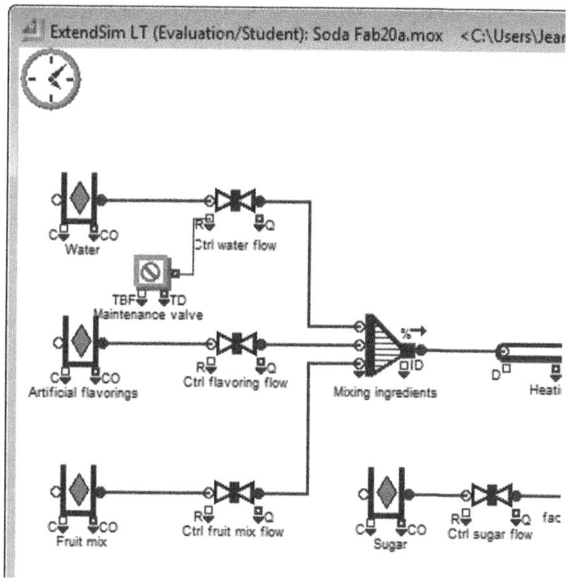

Figure 2.59. *The SHUTDOWN block used to manage the testing of the water valve*

We must then input the parameters in the SHUTDOWNS and OPTIONS tabs of the SHUTDOWN block.

We choose to define a regular check-up of the valve, with an average duration of 15 min plus or minus 5 min, every 4 h. During this time, the valve is shut and there is zero flow through it.

Here are the parameters to define (Figure 2.60):

– selecting FAILURES ARE INDEPENDENT OF REPAIR takes into account a fixed repetition without including the duration of the testing;

– the SET TIME BETWEEN FAILURE (TBF) is set to be a CONSTANT of 4 h, or 240 MINUTES;

– the SET TIME TO REPAIR (TTR) is set to be TRIANGULAR with a duration most likely to be 15 min, but able to vary between a minimum of 10 and a maximum of 20 min;

– the DOWN VALUE is 0 (no flow) and the UP VALUE is 80 (default flow);

– the shift used is the one we named 'Operational'.

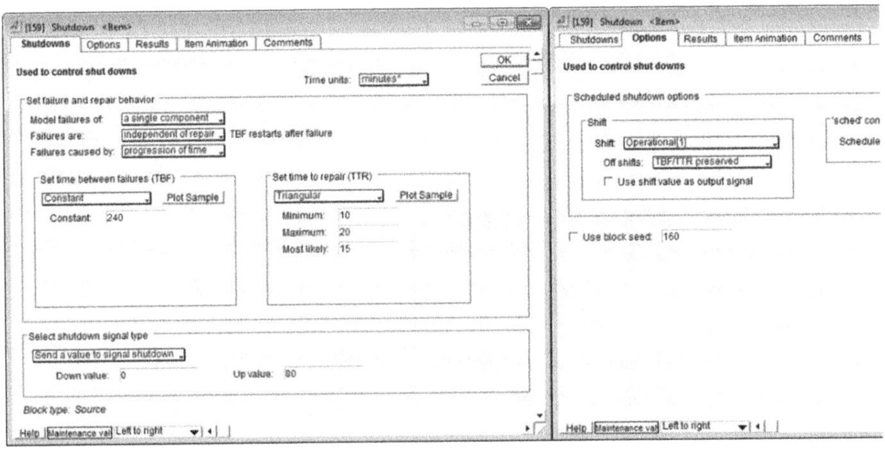

Figure 2.60. *The SHUTDOWN block and the parameters defined in its SHUTDOWNS and OPTIONS tabs*

COMMENT 2.20.– During the simulation, we can see that the valve testing has almost no effect on the production; this is due to the heating conduit which serves as a buffer zone for the mixture (water, flavorings and fruit mix) supply.

2.3.6.1.2. Crating robot

We will also use a SHUTDOWN block for testing the crating robot, which will stop the robot operation by setting its capacity to 0.

The output connector of the 'Maintenance' SHUTDOWN block is connected to the 'C' connector of the 'Crate packaging' ACTIVITY block.

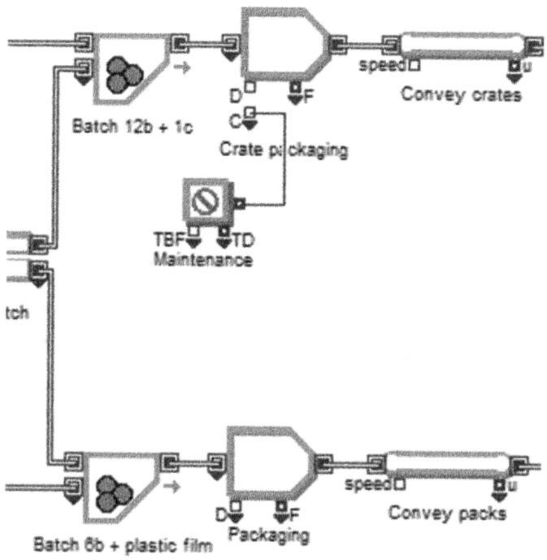

Figure 2.61. *In the center, the SHUTDOWN block connected to the ACTIVITY block*

The frequency and characteristics of the testing are as follows:

– one check-up every 90 min;

– duration: 8 min average with a standard deviation of 2 min;

– schedule: 8:00 am–7:00 pm ('Operational').

Figure 2.62 shows the settings.

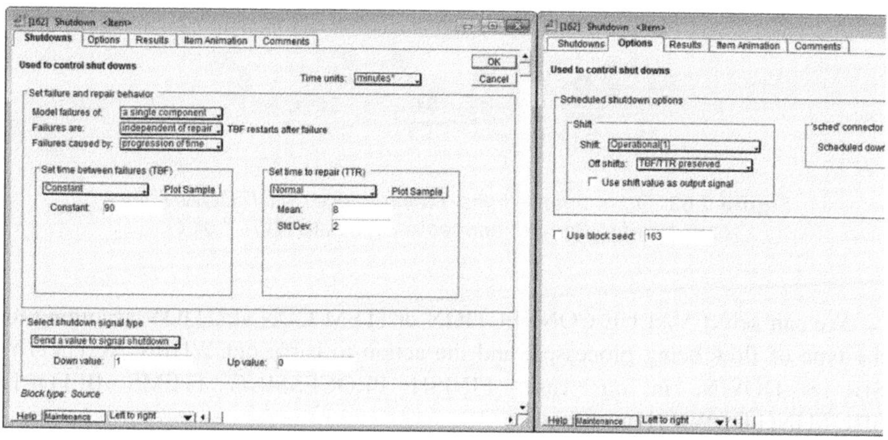

Figure 2.62. *The 'Maintenance' SHUTDOWN block parameters*

There is also another way of programming the testing with an additional method which could be useful.

When a planned stop occurs, it can be that the activity is in the middle of filling a crate, in which case the machine stops anyway without further consideration. In this second solution, we will be able to decide what happens to the crate being filled.

To achieve this, we will configure the 'Crate packaging' ACTIVITY block so that the instruction to stop comes from a specific connector labeled 'SD' as shown in Figure 2.63.

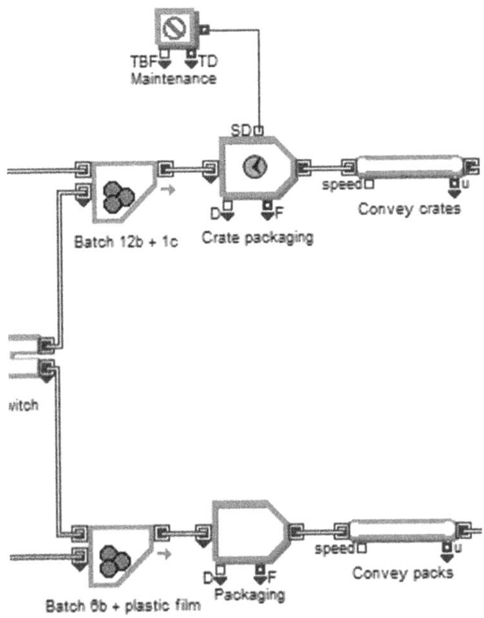

Figure 2.63. *Our model with the 'Maintenance' SHUTDOWN block connected to the 'SD' connector of the ACTIVITY block*

We can select VALUE CONNECTION or ITEM CONNECTION according to the type of flow being processed, and the action to carry out WHEN ACTIVITY SHUTS DOWN, in our case: FINISH PROCESSING ITEMS BEFORE SHUTTING DOWN (Figure 2.64).

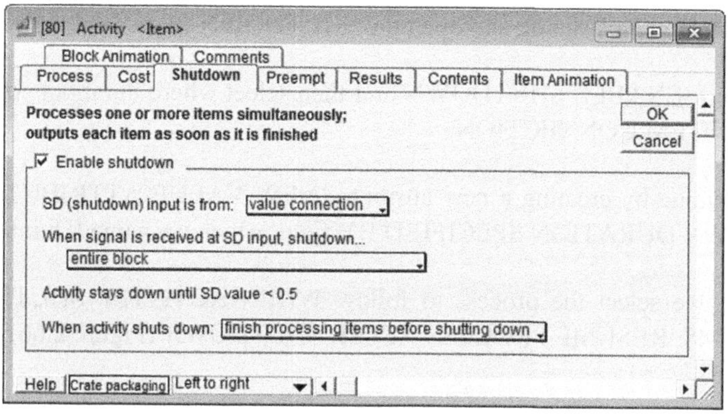

Figure 2.64. *The configuration of the ACTIVITY block under the SHUTDOWN tab*

We have to change the configuration of the 'Maintenance' SHUTDOWN block, under the SHUTDOWN tab, so that it sends the value 1 in the event of a shutdown, and 0 by default, to the 'SD' connector (Figure 2.65).

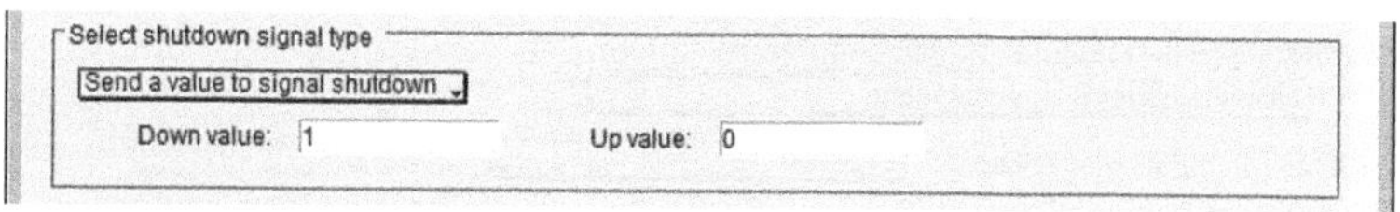

Figure 2.65. *The configuration settings for the shutdown in the bottom part of the SHUTDOWN tab in the SHUTDOWN block*

COMMENT 2.21.–

– When the simulation is in SHOW 2D ANIMATION mode, the ACTIVITY block is covered by a red cross during testing, indicating its stoppage.

– A SHUTDOWN block can be used to generate random incidents (or breakdowns). The distribution of intervals between shutdowns must simply be set to something other than CONSTANT in the configuration of the SHUTDOWN tabs.

2.3.6.1.3. Palletizing robot

In this example, the testing will be programmed according to predefined plans. A CREATE block ('Maintenance planning') will control the ACTIVITY block ('Palletizing') through its 'SD' connector (see the previous section).

To start, we will configure our 'Palletizing' ACTIVITY block in advance so that it is ready to be controlled by the CREATE block that we will add afterward.

We open the block dialog and open the SHUTDOWN tab.

We check ENABLE SHUTDOWN and then select where the input comes from, in this case ITEM CONNECTION.

We continue by creating a new attribute (NEW VALUE ATTRIBUTE) for the SHUTDOWN DURATION SPECIFIED BY field, which we name ‘Duration’.

Finally, we select the process to follow WHEN ACTIVITY SHUTS DOWN, KEEP ITEMS, RESUME PROCESS AFTER SHUTDOWN (Figure 2.66).

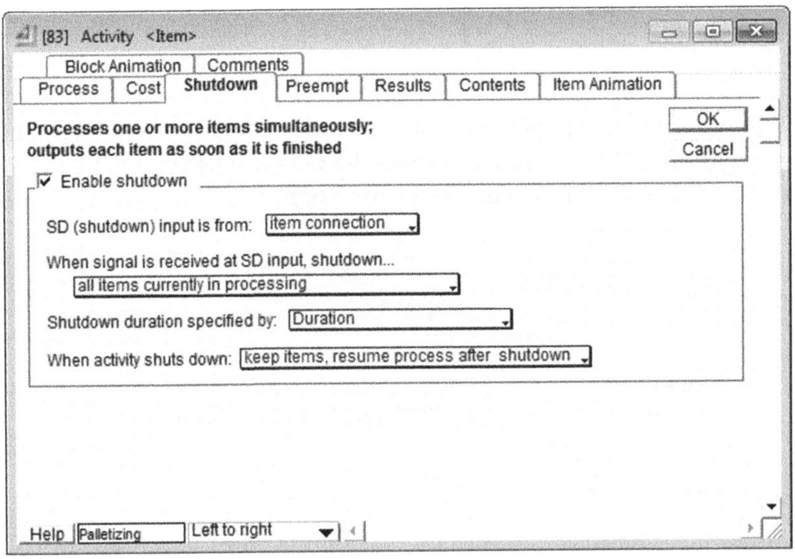

Figure 2.66. *The configuration of the SHUTDOWN tab of the ‘Palletizing’ ACTIVITY tab*

Now, we can add our ‘Maintenance planning’ CREATE block and connect it to the ‘SD’ connector of the ACTIVITY block, see Figure 2.67.

In this block, we will set up a plan supporting two distinct operations:

– minor maintenance at 11:30 am, average duration: 6 min;

– major maintenance at 3:00 pm, average duration: 18 min.

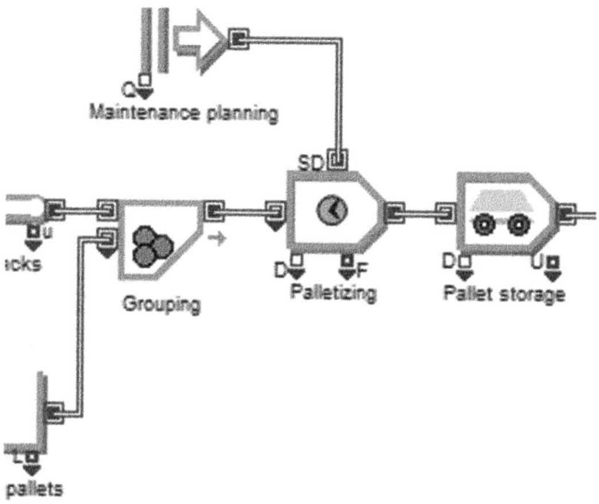

Figure 2.67. *The 'Maintenance planning' CREATE block connected to the 'Palletizing' ACTIVITY block*

To model these characteristics, we configure the CREATE tab (Figure 2.68), under which we set the following:

– the block behavior, here CREATE ITEMS BY SCHEDULE;

– the unit of time, which by default must be MINUTE;

– the number of rows and columns we need (click on the green +/– square), here two rows and four columns;

– the format of CREATE TIMES, here CALENDAR FORMAT;

– the dates and times of the maintenance periods in the CREATE TIME column (3/1/2016 11:30:00 and 3/1/2016 15:00:00);

– their duration, in the fourth column, which also takes the 'Duration' attribute previously created in the 'Palletizing' ACTIVITY block.

COMMENT 2.22.– By default, the unit of time is kept from the overall model, in our case MINUTE (see the RUN menu, SIMULATION SETUP). This is indicated here by an asterisk (*) to the right of the unit.

When we run the simulation, there may be no observable difference – this is not due to our maintenance planning but rather the fact that the testing periods fall during the waiting periods of the robot (when the pallet contains fewer than 72 packs).

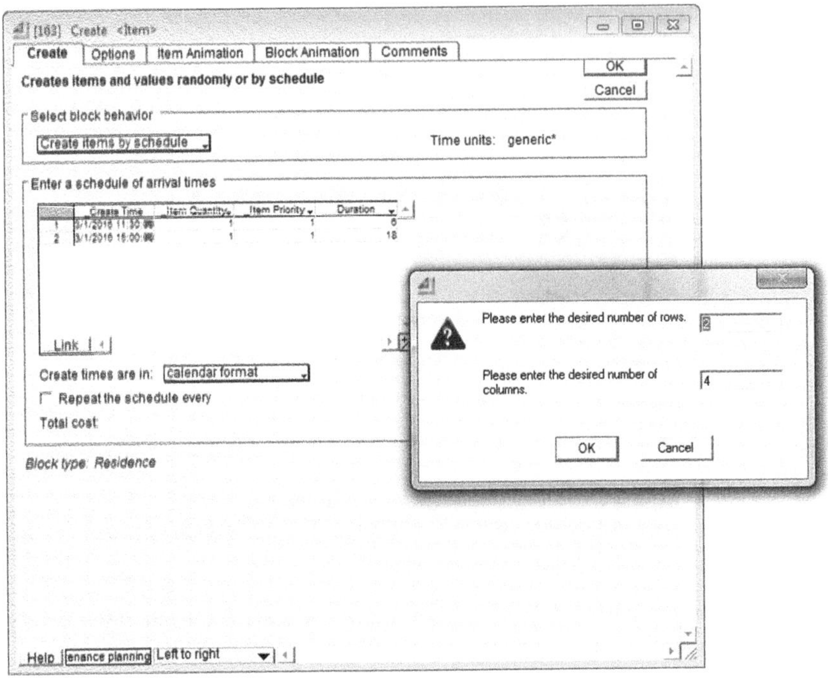

Figure 2.68. *The CREATE tab of the 'Maintenance planning' CREATE block with its calendar and various parameters*

2.3.6.2. *Improving model usability*

In order to simplify the appearance and interface of our model, we can organize the blocks into a hierarchy, as described in section 1.3.9.

The groupings retained are those shown in Figure 2.69, which shows four distinct segments:

– production of the soda up to its bottling;

– transport, carbonation and labeling;

– packaging, transport and storage;

– maintenance and cleaning.

After grouping together and constructing the hierarchical blocks, we should obtain a model resembling that shown in Figure 2.70.

Each of these blocks is represented by a different symbol, only the 'Operational' and 'Bottling' calendars and the STATISTICS block are left outside the hierarchical structure.

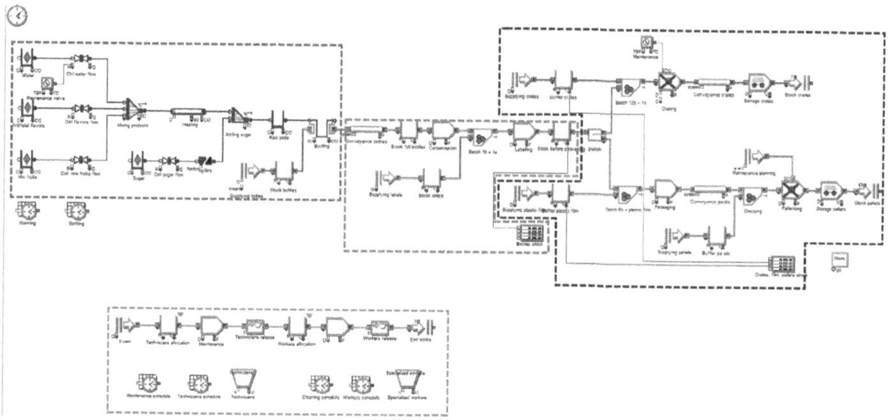

Figure 2.69. *The breakdown of our model into its four hierarchical blocks. For a color version of this figure, see www.iste.co.uk/reveillac/modeling3.zip*

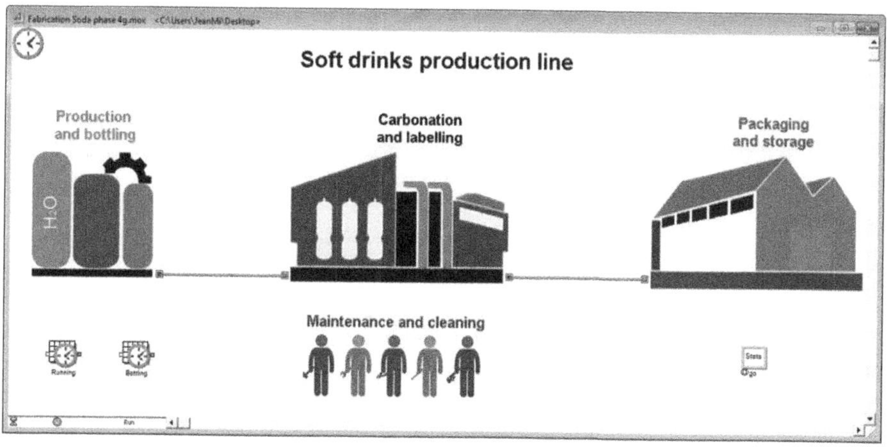

Figure 2.70. *One possible representation of our model with its hierarchical blocks. For a color version of this figure, see www.iste.co.uk/reveillac/modeling3.zip*

Along with the hierarchical blocks, we can add various input zones allowing the supply rates for the bottles, labels, crates, films and pallets to be varied, as well as a control for tuning the speed of the 'Convey bottles' CONVEY ITEM block.

Applying these methods as shown in section 1.3.10, our new model should resemble the one shown in Figure 2.71. We can see the three dialog zones in which the supply rates can be modified and a SLIDER ('Utilities.lix' library) which controls the bottle conveyor speed.

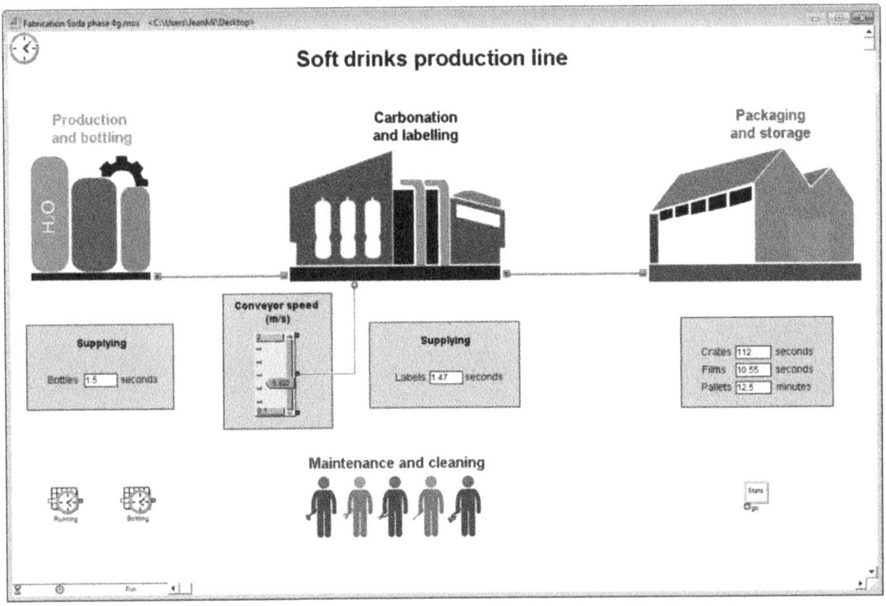

Figure 2.71. *Our hierarchical model with dialog zones and the speed control (SLIDER) for the bottle conveyor. For a color version of this figure, see www.iste.co.uk/reveillac/modeling3.zip*

2.4. Conclusion

ExtendSim offers several options for dealing with mixed flows which are not presented here, such as the ability to generate a 3D animation from a model, to create personalized blocks and to manipulate databases using the Microsoft Excel *add-in*.

If what you have seen in Chapters 1 and 2 has been of interest, you can continue by studying the software guide and working through some of the examples provided.

A demonstration version of the ExtendSim software is downloadable from the publisher's Website (see the Internet links at the end of this book).

3

3D Flows and Evacuation Simulation

3.1. 3D flows

Chapters 1 and 2 considered discrete flows (skiers at a winter sports resort) and mixed flows (the production and bottling of soda), for which 2D simulations were implemented using the ExtendSim software.

We will now turn our attention to flows for which 3D representation is of utmost importance. This is the case for evacuations of buildings or public areas or even airplanes, trains or subway carriages, as well as for transporting components on a production line, for simulating road or pedestrian traffic or for planning the crossroads or connections in a town's traffic, bus or tramway routes, for example.

3D treatment is required because the volumes of the obstacles and free spaces involved, and the layouts of the places considered, are important elements of the simulation. For example, it is clear that the volume of a package will influence the dimensions of the conveyor belt that transports it, that the evacuation of a multi-story building must take account of the stairs or elevators, that a person cannot step over his/her desk or walk through walls, that a vehicle is confined to the road network or that a tramway depends upon the rail infrastructure as determined by the local geography.

In this chapter, we will start by presenting and working through an exercise involving the evacuation of a building using the Pathfinder software developed by Thunderhead, after which we will move onto the FlexSim software in Chapter 4.

3.2. The Pathfinder software

This is a specialist piece of software for simulating evacuations. It is able to import files in the DXF and DWG (AutoCAD) formats, as well as images in JPG, PNG or GIF format for use as backgrounds to ease the design of 2D or 3D models. Evidently, the flows modeled in this application are composed entirely of people.

The 3D representation is based upon a triangulated mesh that is able to reproduce complex geometric features with high precision.

Pathfinder can run its simulations in one of two modes, a standard mode and a mode which follows the guidelines of the SFPE (Society of Fire Protection Engineers) incorporating, among other things, the concepts of people density, speeds and movement limits.

The people or occupants being simulated in the software exhibit completely configurable behavior as a result of fully editable profiles. They are designed according to multiple characteristics such as age, clothing or profession.

Several tens of thousands of people can be animated in real time.

During the simulation, various types of data visualization can be overlaid over the model, such as occupation density, speeds or times.

Elevators, escalators, ramps and moving walkways, along with their properties, can be easily introduced into the 3D model.

Flow restrictions such as queues, doors or compulsory routes can be combined to manage occupant flows in relation to the specific circumstances.

Data are generated in graphical or spreadsheet-compatible CSV formats, allowing precise analysis of all relevant parameters (journey time, duration of the evacuation, people numbers, etc.).

Furthermore, software like PyroSim (which simulates fire dynamics and tests smoke distribution) can be used in combination with Pathfinder.

Figure 3.1 gives an overview of Pathfinder after start-up.

A free, completely operational demonstration version of the software, limited to 30 days' use after installation, is available for download from: www.thunder

headeng.com/pathfinder/download/ or by clicking on FREE TRIAL at www.thunder headeng.com/pathfinder/[1].

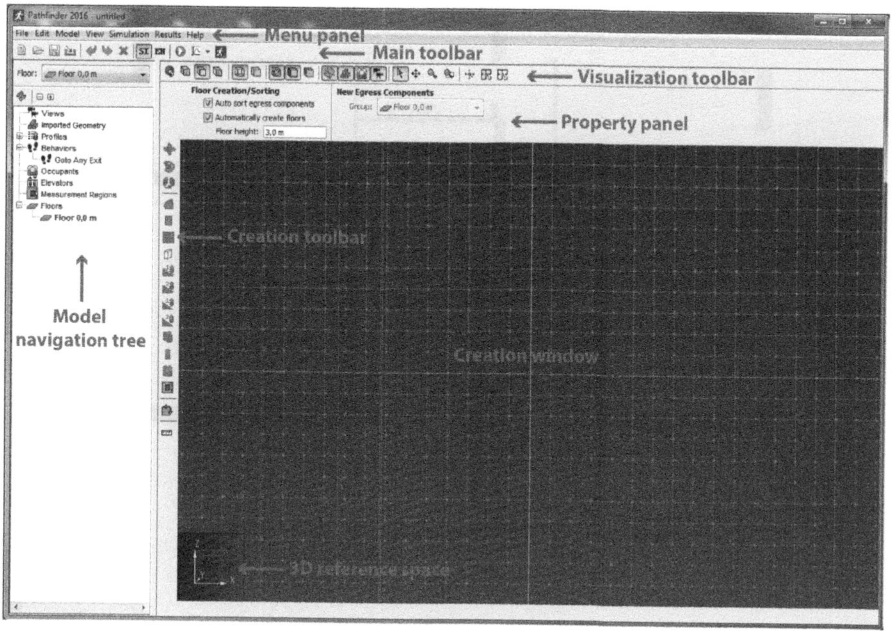

Figure 3.1. *The different windows and toolbars in the Pathfinder software*

3.3. Evacuation of a building with PathFinder

For this example, we will import the plans of the building as JPEG images exported from the 2D plans, one for the first floor and another for the second floor.

On the first floor, we can see the double-door main entrance at the bottom of the plan as well as the service entrance, situated on the left, between the administration office and the meeting room. The building is 19.20 m across and 13.20 m deep. A stairway with two flights of stairs separated by a landing leads to the second floor. The height between the floors is 3 m.

1 This offer is available at the time of writing but cannot be guaranteed indefinitely, although it has been available for several years with periodically updated versions.

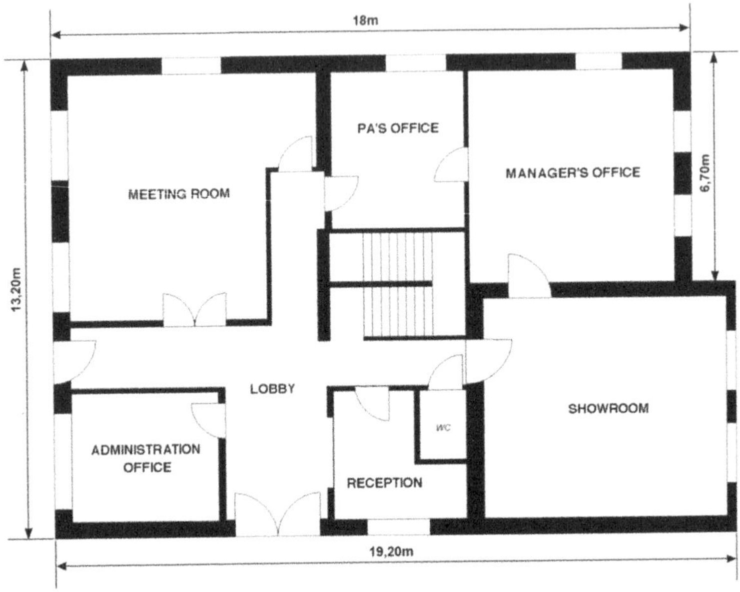

Figure 3.2. *Plans for the first floor of the building*

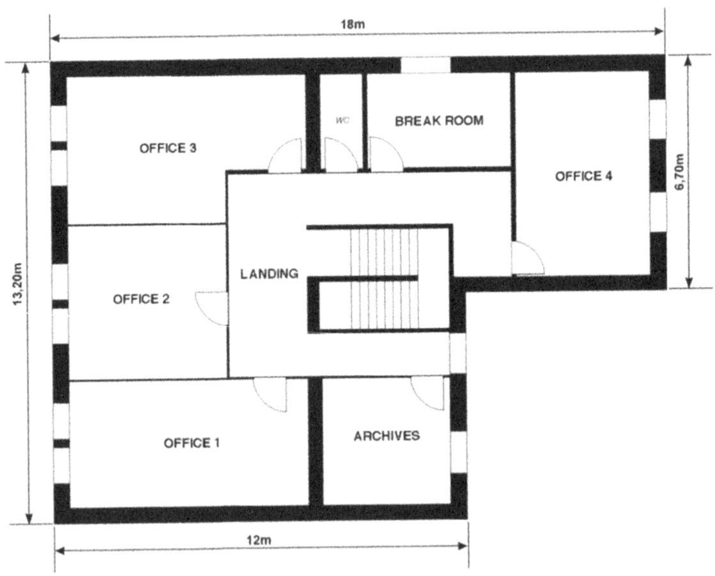

Figure 3.3. *The second floor of the building*

The second floor does not extend over the full surface area of the first floor. The height up to the ceiling is set to 3 m.

This simulation will be constructed in several steps to familiarize you with the software:

- importing and formatting the first floor plans;
- creating the different first floor rooms;
- creating the first floor doors;
- populating with occupants;
- simulation and results for first floor evacuation;
- incorporating furniture;
- importing and formatting the second floor plans;
- creating the rooms and doors, and populating with occupants;
- creating the stairs;
- simulation and results for evacuation of the whole building.

3.3.1. *Importing and formatting the first floor plans*

After launching Pathfinder, we specify that we will use the International System of Units (SI) by selecting its symbol from the main toolbar.

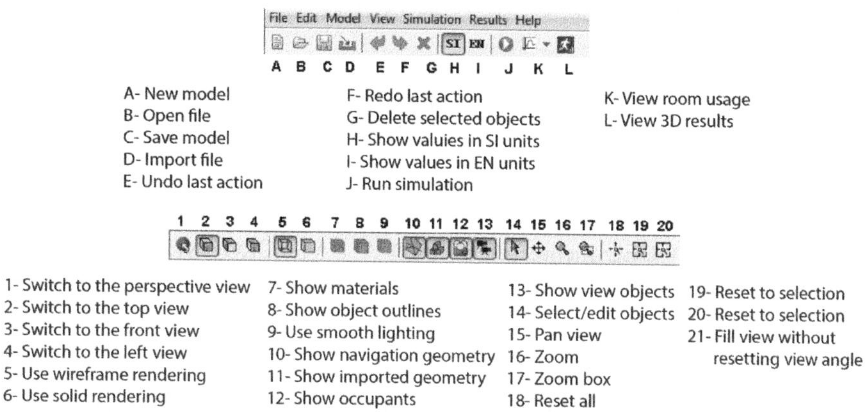

Figure 3.4. *The two horizontal toolbars in Pathfinder. The choice of 'SI' or 'EN' units can be seen (above), along with the other tools*

We continue by setting the spacing of the navigation grid. Go into the VIEW menu, select EDIT SNAP GRID… and input a spacing of 0.2 in the SNAP GRID SPACING field. Inputting units (m) is not required.

If the correct visualization options are selected, in particular SWITCH TO THE TOP VIEW in the main window toolbar, you should see the grid appear as the background in the main window (Figure 3.5).

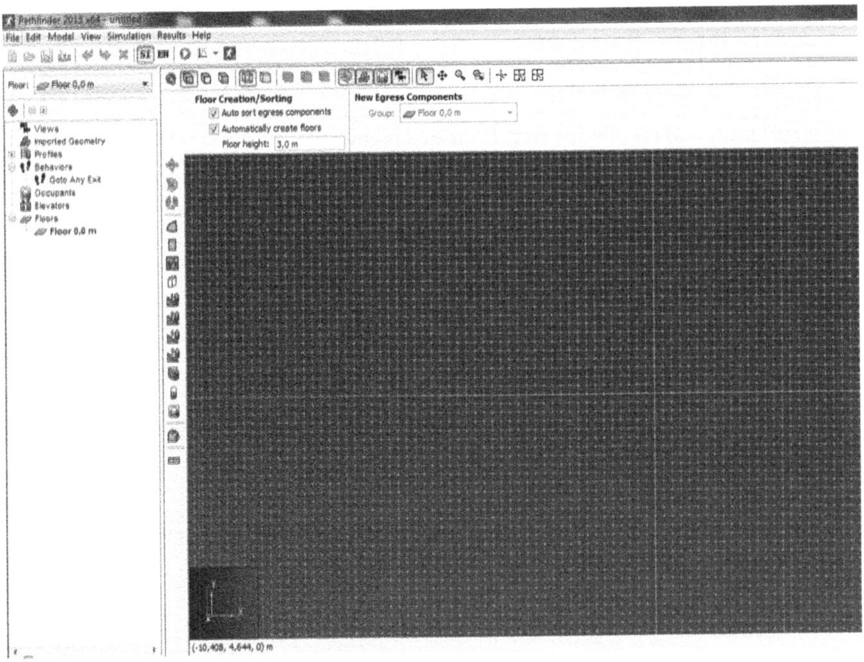

Figure 3.5. *The Pathfinder visualization grid in top-down view (the second symbol in Figure 3.4)*

We now import the plans for the first floor of our building.

Go into the MODEL menu, select ADD A BACKGROUND IMAGE… and choose your plans. The BMP, GIF, JPG, JPEG, PNG, TIF, TGA, TARGA, JIF and JPE formats are supported.

Click on the OPEN button to begin importing. A new window called CONFIGURE BACKGROUND IMAGE opens. From here, we can set up how our plans are used as the background image.

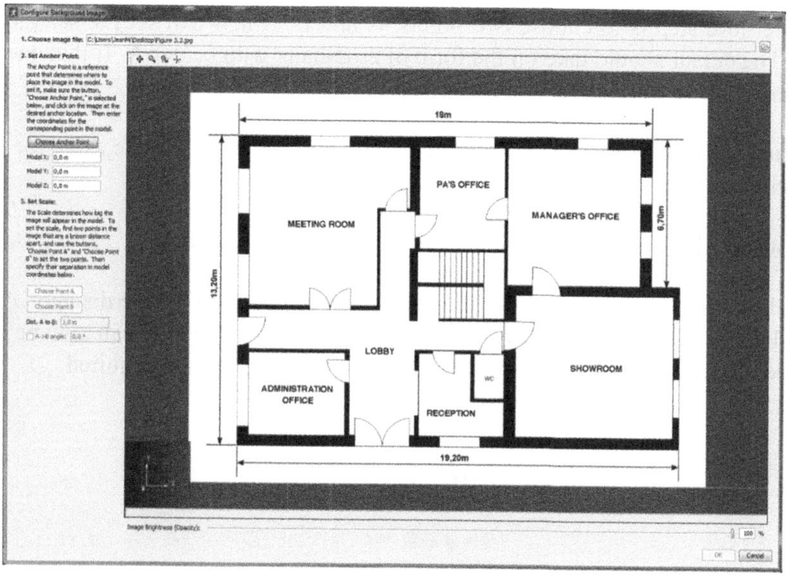

Figure 3.6. *The configuration window for the background image*

The top field shows the selected image file. We must now set the anchor point using the blue crosshairs labeled ANCHOR. If these crosshairs do not appear, click on the CHOOSE ANCHOR POINT button.

The anchor point determines the (X, Y, Z) origin coordinates, for example the bottom left-hand corner of our plans (Figure 3.7).

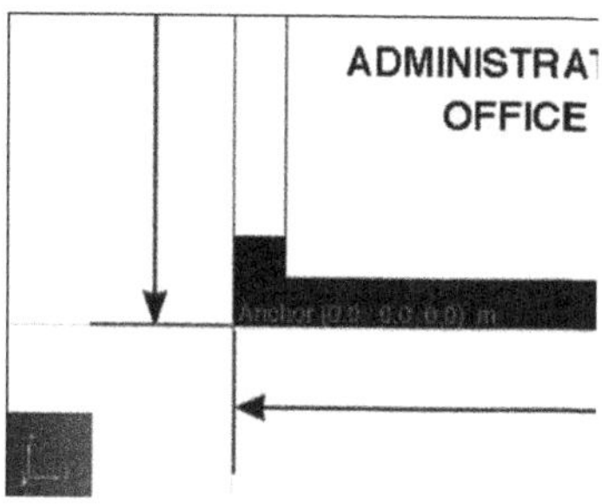

Figure 3.7. *Establishing the anchor point. For a color version of this figure, see www.iste.co.uk/reveillac/modeling3.zip*

One click positions the marker. The zoom functions in the top toolbar allow for greater precision.

The third step is to choose two points, A and B, which will set the scale of the plan. The cursor should take on the form of crosshairs with center A. If this does not happen, click on the CHOOSE POINT A button.

Place the marker for A at the required point, click, and then do the same for point B after clicking on CHOOSE POINT B.

COMMENT 3.1.– Point A can be placed over the origin if required.

All that is left now is to enter the real length represented by the distance AB in the DIST. A TO B field (Figure 3.8). The angle between A and B can also be set if desired, under A->B ANGLE. Inputting the units (m and °) is not required.

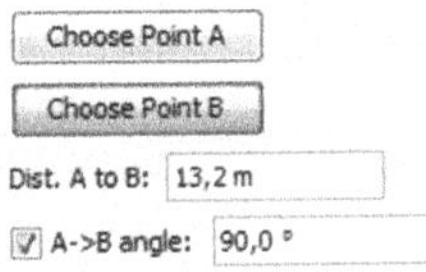

Figure 3.8. *Inputting the distance AB and the angle AB*

COMMENT 3.2.– A slider, labeled IMAGE BRIGHTNESS (OPACITY), is present at the bottom of the window. This allows the opacity or transparency of the background image to be set from 0 to 100% as required.

Click OK to confirm the parameters and make your image appear in Pathfinder's creation window.

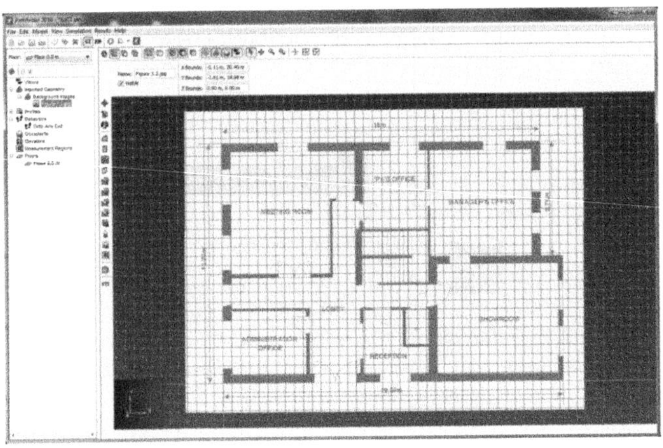

Figure 3.9. *Pathfinder with the first floor plan as background image*

COMMENT 3.3.– The icon at the bottom-left of the main window shows the orientation of the X, Y and Z axes relative to the selected point of view (top-down, face-on, perspective or side-on).

3.3.2. *Creating the different first floor rooms*

The creation toolbar has to be used in order to create the different rooms to be overlaid on the plan.

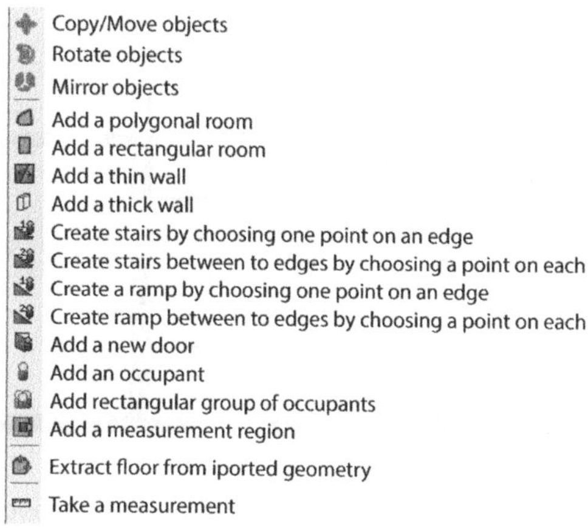

Figure 3.10. *The different symbols on the creation toolbar*

Most of these tools have two modes of action – normal mode (accessed *via* a single click on the icon) used for a single action, for example the creation of a rectangular room, and 'sticky' mode (accessed *via* a double-click) for implementing a sequence of actions, such as the creation of several rooms, one after the other. After double-clicking, a green dot appears above the selected icon.

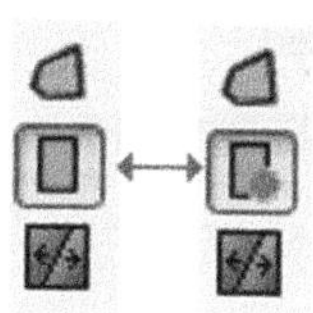

Figure 3.11. *The ADD A RECTANGULAR ROOM tool in normal mode (on the left) and 'sticky' mode, with the green dot (on the right)*

For our example, as is often the case, we will use the ADD A RECTANGULAR ROOM tool in 'sticky' mode and draw a series of rectangles over the different parts making up the first floor rooms as shown in Figure 3.12.

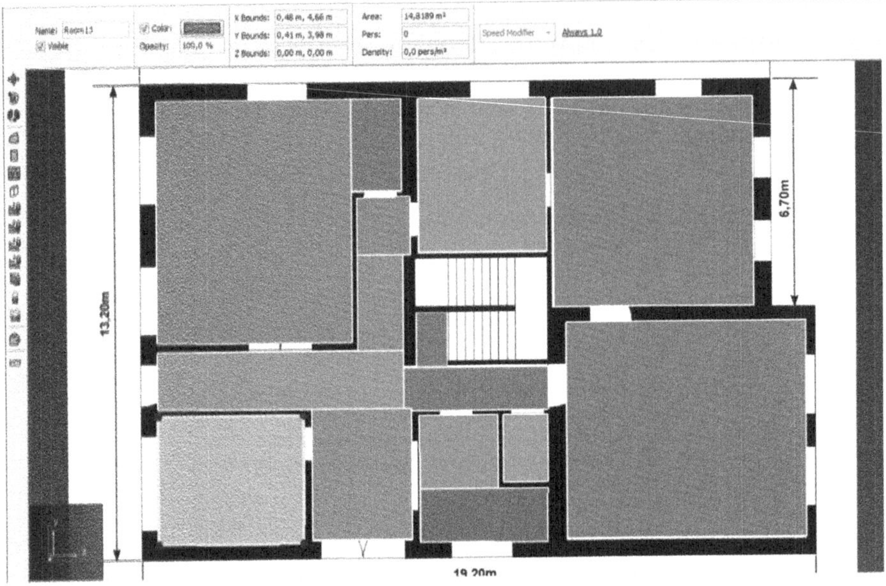

Figure 3.12. *The different rectangles outlining the first floor rooms. For a color version of this figure, see www.iste.co.uk/reveillac/modeling3.zip*

After being drawn, the rectangles added using the 'Add a rectangular room' tool appear in yellow, with four blue dots in their corners. These four dots can be used to redefine the size of the rectangle if required, using the select/edit tool, number 14 in Figure 3.4 (the one that looks like a pointer).

COMMENT 3.4.–

– For greater precision when drawing, the pan view (number 15) and zoom (numbers 16 and 17) tools can be used.

– The fitting/framing tools (numbers 18, 19 and 20) may also be needed.

– To delete a rectangle, simply select the rectangle and press the DELETE key on the keyboard.

– If the grid annoys you, you can get rid of it by deselecting the SHOW SNAP GRID option in the VIEW menu. In this case, the 'snapping' feature disappears and the room boundaries are no longer confined to the grid.

– After drawing a rectangle, we can input an opacity percentage between 0 and 100% in the OPACITY field in the property panel (see Figure 3.1) to have a better view of the rooms in the plan.

– The color of the different rectangles can be modified using the COLOR option from the property panel.

Together, the rectangles that we have drawn define all of the room boundaries, but we still need to merge some of them to exactly replicate the floor plan.

To do this, simply select the relevant rectangles while holding down the CTRL key on the keyboard and then right-click before selecting the MERGE option (or use CTRL M).

You should obtain a model resembling the one shown in Figure 3.13.

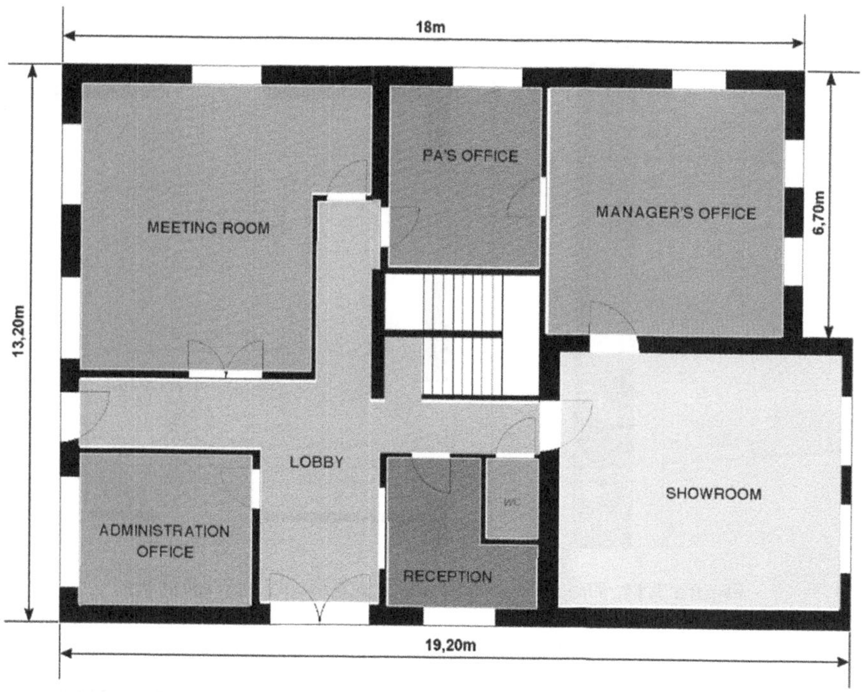

Figure 3.13. *Our plan after merging all of the relevant rectangles to create the different rooms present on the first floor. The opacity of the rectangles has been set to 65% to allow the names to be visible, and the colors have been changed by default for better visual effect. For a color version of this figure, see www.iste.co.uk/reveillac/modeling3.zip*

COMMENT 3.5.– The rooms could have been drawn using the ADD A POLYGONAL ROOM tool, which would have avoided the rectangular divisions and hence the merging afterward, however, this tool is arguably less precise and slower to use.

The different rooms are shown in the navigation tree on the left, under the FLOORS branch, then FLOOR 0.0 M (the default first floor elevation). By selecting the rooms one-by-one, they can be renamed from the property panel.

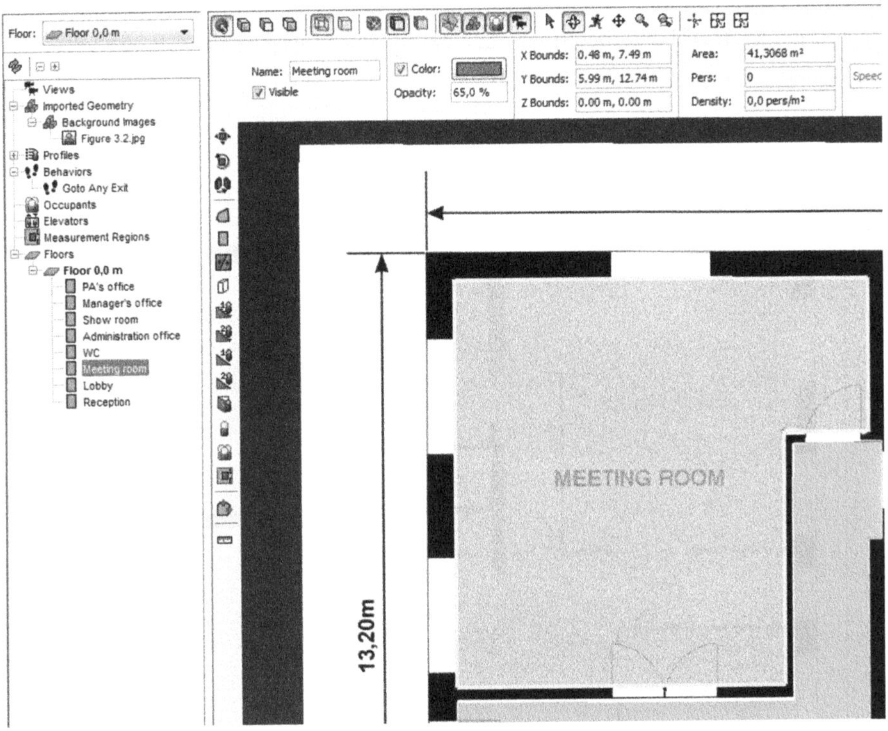

Figure 3.14. *The navigation tree for our model with all of the created and renamed rooms. The name field ('Meeting room') is visible in the property panel, as well as the corresponding selected room*

3.3.3. *Creating the first floor doors*

We will install two types of doors: interior doors, between rooms, and exterior doors, opening to the outside.

Select the ADD A NEW DOOR tool (see Figure 3.10) and create the different doors between each of the rooms by drawing a rectangular zone corresponding to the position and dimensions of each of the doors, then the two exterior doors as shown in Figure 3.15.

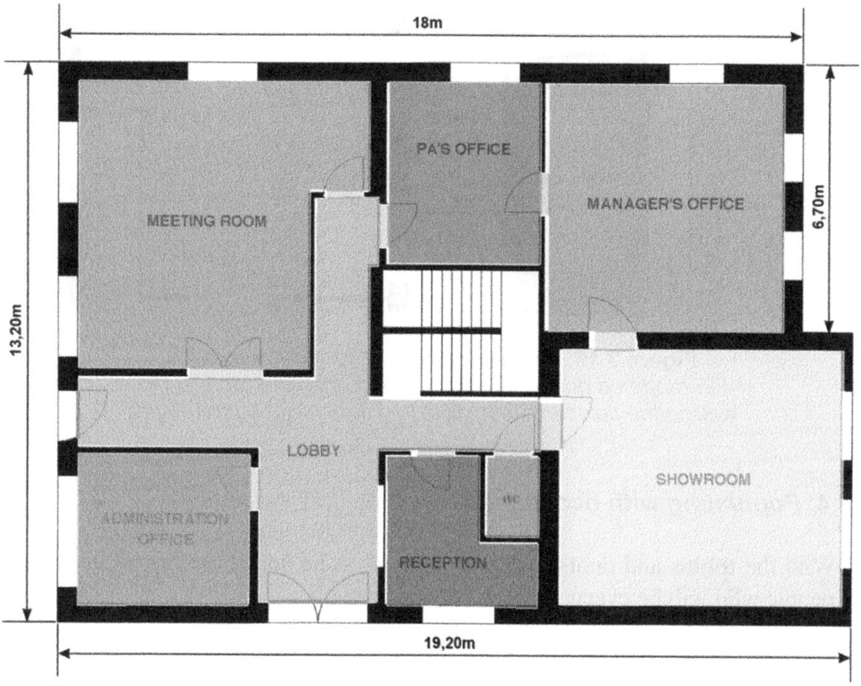

Figure 3.15. *Our model with all of its interior doors (in yellow) and the two exterior doors (in green). For a color version of this figure, see www.iste.co.uk/reveillac/modeling3.zip*

Precise door widths can be set by changing the value in the WIDTH field on the property panel. By default, the units used are centimeter but this does not have to be input.

COMMENT 3.6.– The thickness of doors leading outside cannot be adjusted.

As for the rooms, the name of each of the doors can be input in the NAME field in the property panel by selecting each of them from the model's navigation tree.

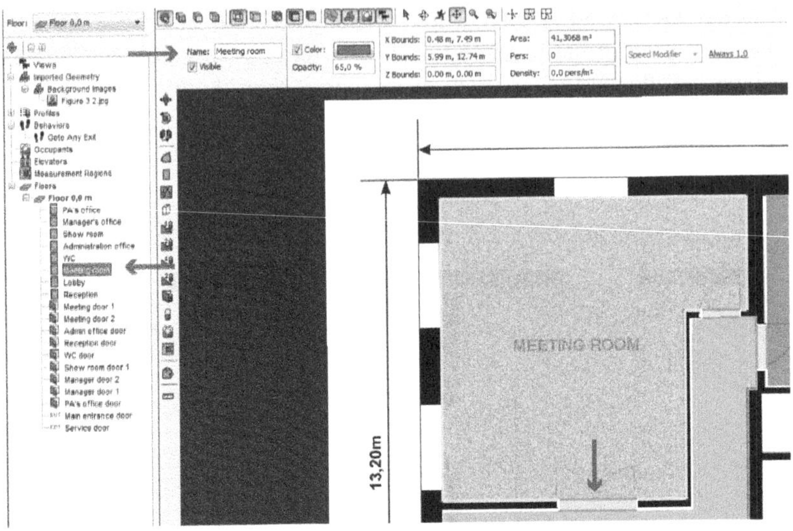

Figure 3.16. *The navigation tree for our model with the renamed doors. The NAME field is visible (above), as well as the selected door, 'Meeting door 1' (at bottom-right)*

3.3.4. *Populating with occupants*

With the rooms and doors in place, the next to be added are the occupants, i.e. the people who will be evacuating the building.

To start with, we will do some testing with the first floor.

Click on ADD AN OCCUPANT (Figure 3.10) and place several of them into the model, as shown in Figure 3.17.

The occupants can be renamed by selecting them from the navigation tree and using the RENAME option that appears after right-clicking, or by selecting them with the pointer tool and inputting the name in the NAME field on the property panel.

COMMENT 3.7.–

– The color of each occupant can be changed using the COLOR check-box and color choice zone in the property panel. This can be useful for improved visualization and monitoring of people's movement during the simulation.

– For the moment, by default, the occupants' behavior and profile are independent of their color. These parameters will be refined later on in this example.

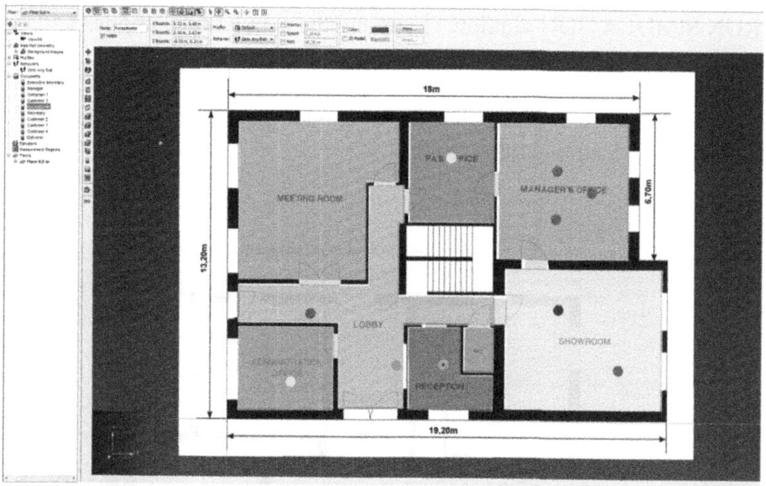

Figure 3.17. *The occupants placed in our model. Their names appear under the OCCUPANTS branch in the navigation tree. Each category of occupants has a different associated color. For a color version of this figure, see www.iste.co.uk/reveillac/modeling3.zip*

3.3.5. *Simulation and results for the first floor evacuation*

With all of the required components in place for the first floor, the evacuation simulation can be run to check that everything works correctly.

Go to the main toolbar and click on RUN SIMULATION (see Figure 3.4).

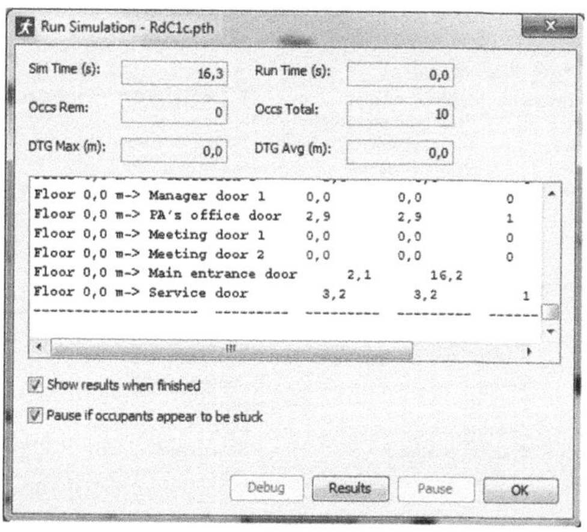

Figure 3.18. *The Run Simulation dialog window*

A RUN SIMULATION window opens and the simulation compiles and runs.

If everything works, a new window opens in the main window. This is the simulation window, in which all of the model parameters can be animated, visualized and saved, among other things.

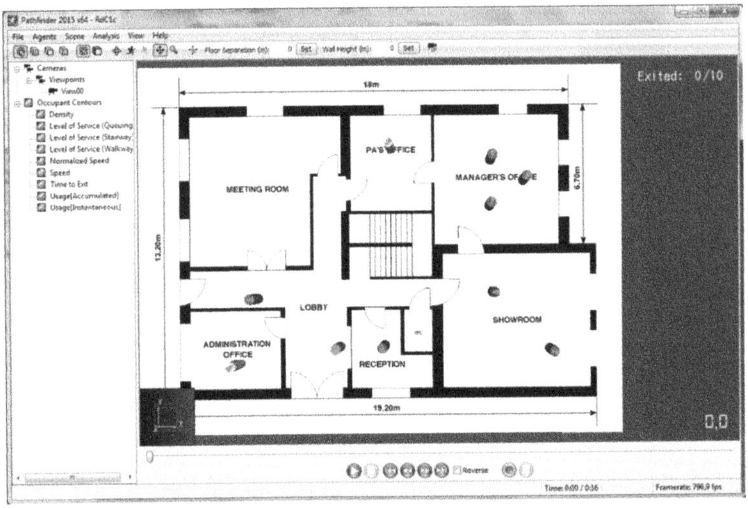

Figure 3.19. *The simulation window for our model*

There are two toolbars in this window, one above and one below, as well as a new tree on the left-hand side.

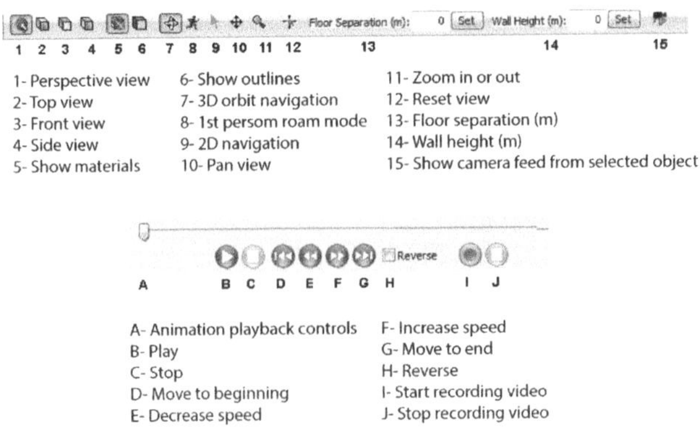

Figure 3.20. *The two simulation window toolbars*

We will run the simulation with the occupants in 'avatar' mode, using a viewpoint like the one shown in Figure 3.21.

Go to the AGENTS menu, select SHOW AS PEOPLE and then the 3D ORBIT NAVIGATION tool (see Figure 3.20) and tilt the model using the mouse. Use the ZOOM and PAN VIEW tools to optimize the plan visualization.

COMMENT 3.8.– When the 3D ORBIT NAVIGATION tool is selected, holding down the SHIFT key allows you to pan the camera, while holding down the ALT key provides a zoom in and out. The scroll wheel on the mouse can also be used to zoom in and out of a specific point.

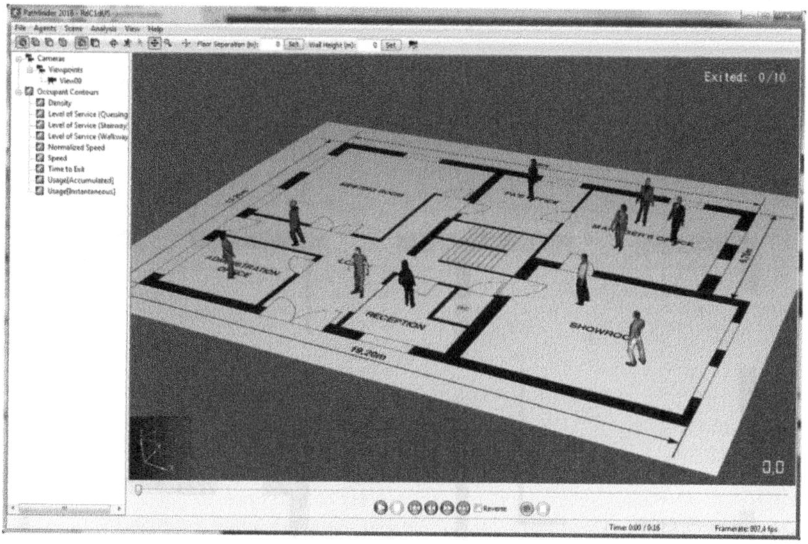

Figure 3.21. *Our model, with its occupants in 'avatar' mode, correctly placed in the middle of the visualization window*

Next click on PLAY to launch the simulation.

You should see the occupants move and leave the building. This action should last around 16 s. The time passed is visible at the bottom-right of the simulation window.

The animation shows that almost everyone leaves *via* the main door ('Main entrance door'). Only one visitor located in the hall, very close to the service entrance ('Service door'), leaves using this route.

This simulation uses the default evacuation logic because none of the occupants' behaviors or profiles have been set.

COMMENT 3.9.– The 3D avatars have not been chosen. The simulation hence shows a default selection of avatars, which may not correspond to the situation. This issue will be addressed and resolved later.

3.3.6. *Incorporating furniture*

During the evacuation, the occupants must move between the different pieces of furniture (desks, wardrobes, chairs, bookcases, tables, seats, etc.) that can get in their way.

Hence, furniture has been added to each of the first floor rooms, as shown in Figure 3.22.

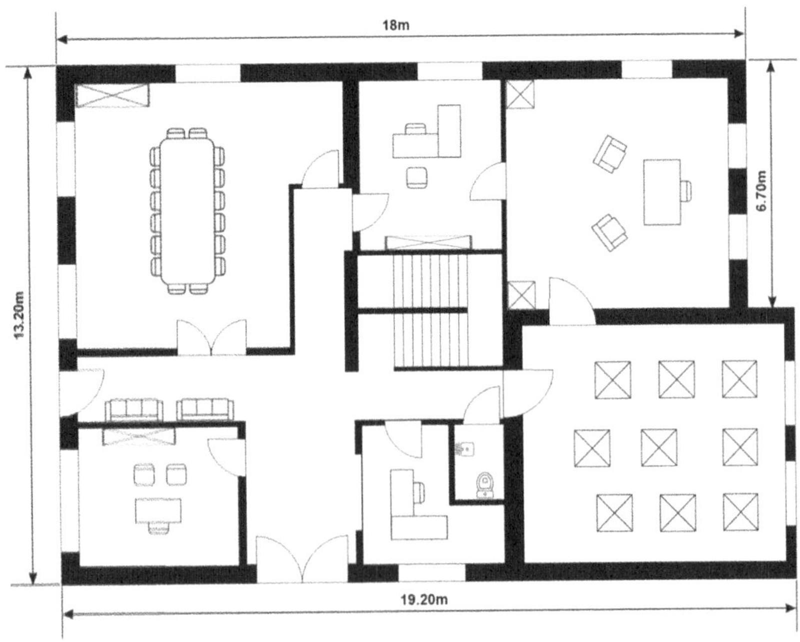

Figure 3.22. *The first floor plan with the added furniture*

After overlaying a new image showing the furniture onto the existing background image (MODEL menu, then ADD A BACKGROUND IMAGE...), you simply trace around each of the pieces of furniture with the ADD A POLYGONAL ROOM or ADD A RECTANGULAR ROOM tools, then delete them by pressing the DELETE key on the keyboard.

Every time you delete something, a dialog box opens asking you to confirm the removal of the outlined object.

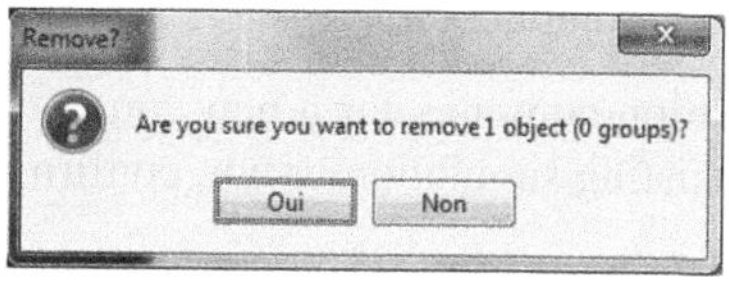

Figure 3.23. *The dialog box confirming the choice to delete an object*

Once all of the furniture has been outlined and removed, you should have a model similar to the one shown in Figure 3.24.

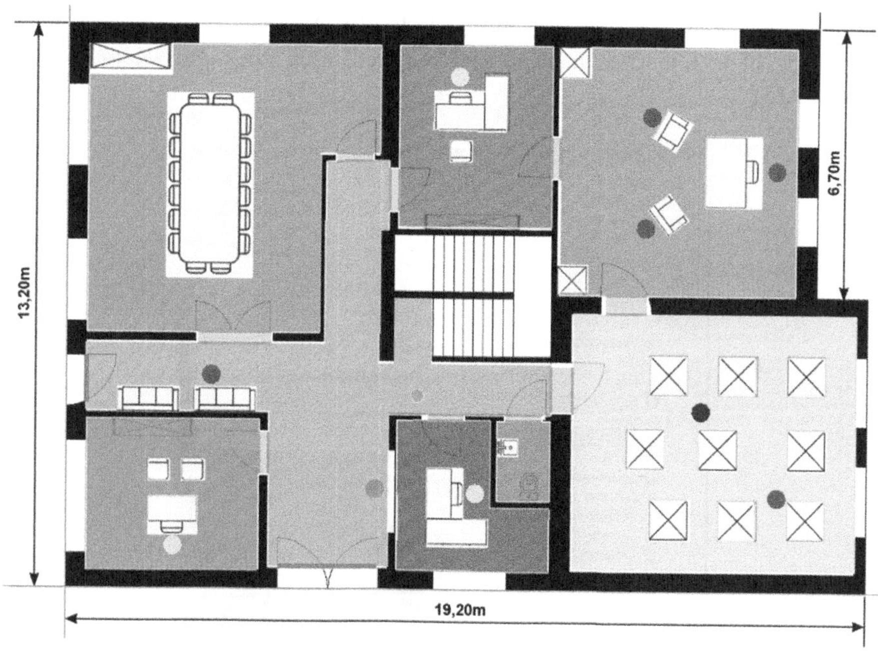

Figure 3.24. *Our model with portions removed from the rooms' accessible areas corresponding to furniture. For a color version of this figure, see www.iste.co.uk/reveillac/modeling3.zip*

If you rerun the simulation, you should see that the occupants avoid the furniture and take slightly longer (around 1 s more) to evacuate the building.

COMMENT 3.10.–

– The occupants must not be placed within the zones removed for the furniture, otherwise they will also be deleted from the model. Hence, care must be taken to move occupants beforehand, if applicable.

– Each time the simulation compiles for a new run, a dialog box appears asking if you want to delete the existing simulation or not, confirm this by clicking YES.

3.3.7. *Importing and formatting the second floor plans*

As specified in section 3.3, the height between the floors is 3 m, so we will create a new floor at this height.

Scroll down the FLOOR list above the navigation tree and select ADD NEW. In the dialog box that opens, next to the ENTER FLOOR LOCATION option, input the height, here 3 m (the units are not required), and click on OK to confirm (Figure 3.25).

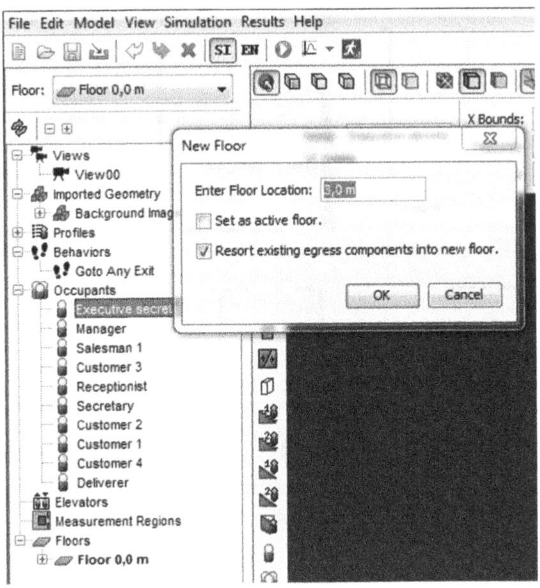

Figure 3.25. *Creating the second floor, 3 m above the first floor*

Next, place yourself on this level by choosing FLOOR: FLOOR 3.0 M from the drop-down list.

3.3.8. *Creating rooms, doors and populating with occupants*

We must now import the second part of the plans, corresponding to the second floor plan, as a new background image at the height of this floor (3 m).

Go to the MODEL menu, select ADD A BACKGROUND IMAGE... and choose the relevant image. As in section 3.3.1, we designate the anchor point, then points A and B separated by 13.20 m at 90°.

Our model should resemble the one shown in Figure 3.26.

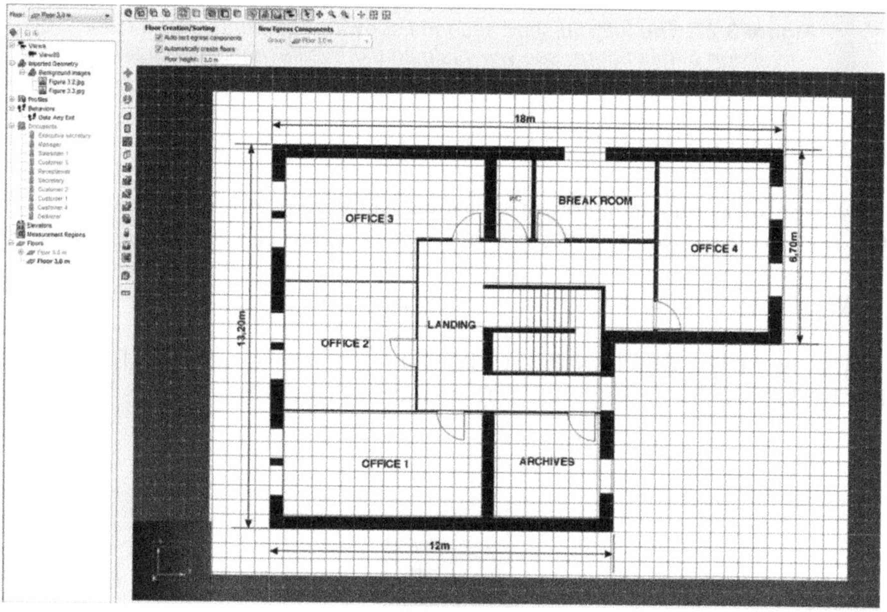

Figure 3.26. *The second floor plan set as the background image at a height of 3 m*

The subsequent operations to be carried out are identical to those in sections 3.3.2, 3.3.3 and 3.3.4.

Once complete, we should have a model comprising the rooms, the doors and the occupants (seven in total) as in Figure 3.27.

Figure 3.27. *The fully finished second floor with occupants. For a color version of this figure, see www.iste.co.uk/reveillac/modeling3.zip*

3.3.9. *Creating the stairs*

We use stairs with an intermediate landing halfway up, at 1.50 m, to connect the first floor to the second floor. We will thus have to set this up in the model too.

We start by creating the intermediate landing.

Use the ADD A RECTANGULAR ROOM tool and input its height, 1.5, in the Z PLANE field on the property panel (the units are not required).

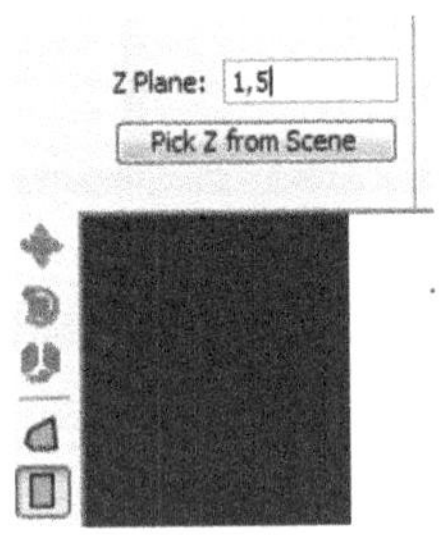

Figure 3.28. *The ADD A RECTANGULAR ROOM tool and the height of 1.5 input in he Z PLANE field*

Next, we must draw the landing, which we can name 'Landing' in the NAME field, onto the model.

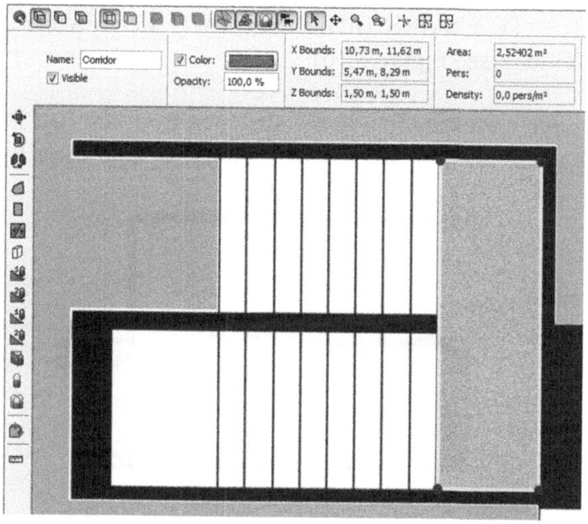

Figure 3.29. *The intermediate landing, after being positioned and named 'Landing' in the NAME field*

We will now display the two stories by selecting <SHOW ALL> from the FLOOR drop-down list above the navigation tree.

By making the rendering of the second floor rooms slightly transparent, you should be able to make out some of the components of the first floor. The showroom should be visible because no room sits above it on the second floor.

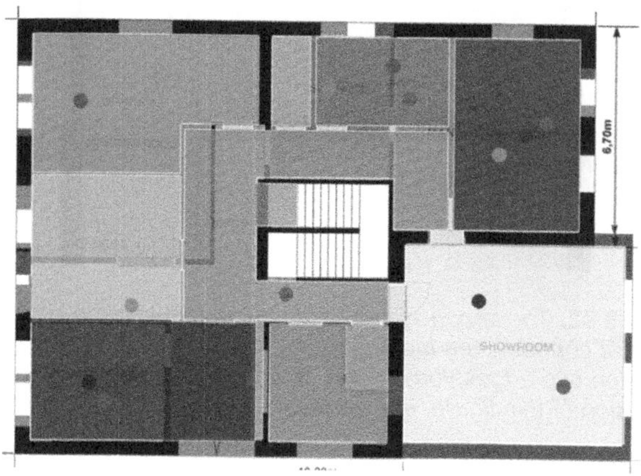

Figure 3.30. *The two first and second floors of the model, displayed simultaneously (<SHOW ALL>). For a color version of this figure, see www.iste.co.uk/reveillac/modeling3.zip*

Select the CREATE STAIRS BETWEEN TWO EDGES BY CHOOSING A POINT ON EACH tool (see Figure 3.10).

Click on the edge corresponding to the first step from the first floor, then on the edge of the landing as shown in 3.31. Stairs should then take shape there.

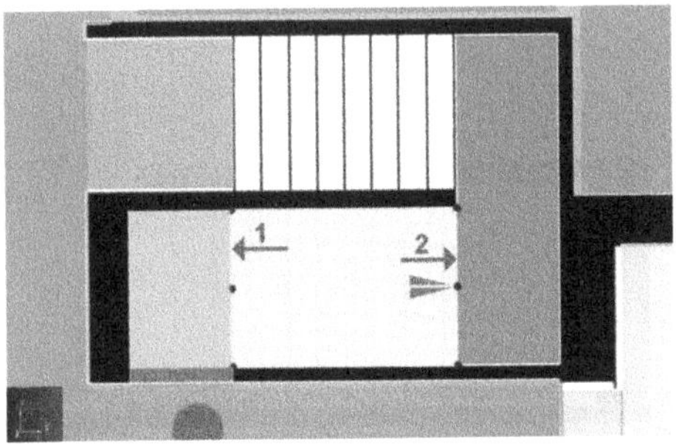

Figure 3.31. *The first flight of steps forming the part of the stairs between the first floor and the landing. The red arrows indicate the two edges you must click on to generate the steps. For a color version of this figure, see www.iste.co.uk/reveillac/modeling3.zip*

Repeat this operation, this time with the landing and the edge of the second floor landing, as in Figure 3.32. A second flight of steps should appear.

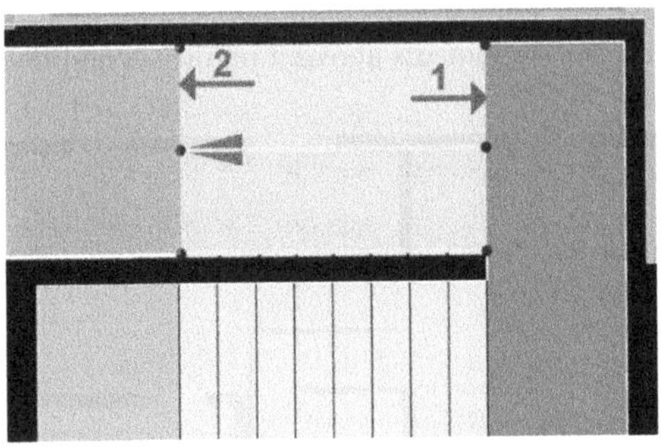

Figure 3.32. *The second flight of steps constituting the part of the stairs that connects the landing to the second floor. The red arrows indicate the two edges you must click on to generate the steps. For a color version of this figure, see www.iste.co.uk/reveillac/modeling3.zip*

If the width of the steps does not cover the width of the stairwell, you can input its value, here 130 cm, in the field next to the WIDTH option from the property panel after selecting a flight of steps (the units are not required).

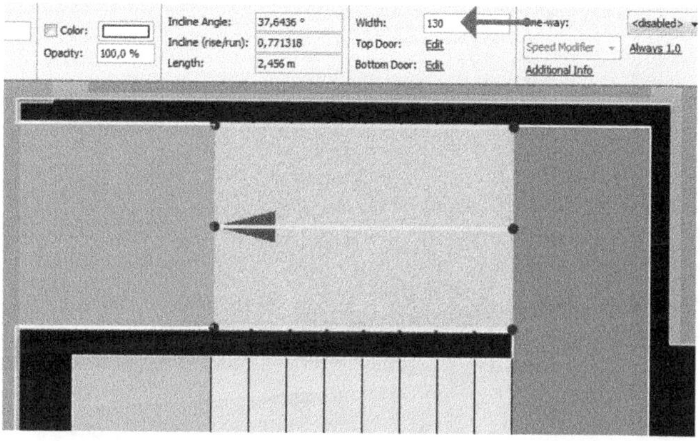

Figure 3.33. *Inputting the step width, here 130 cm, in the WIDTH field*

We can visualize the stairs by clicking on the SWITCH TO THE FRONT VIEW symbol from the visualization toolbar (see Figure 3.4).

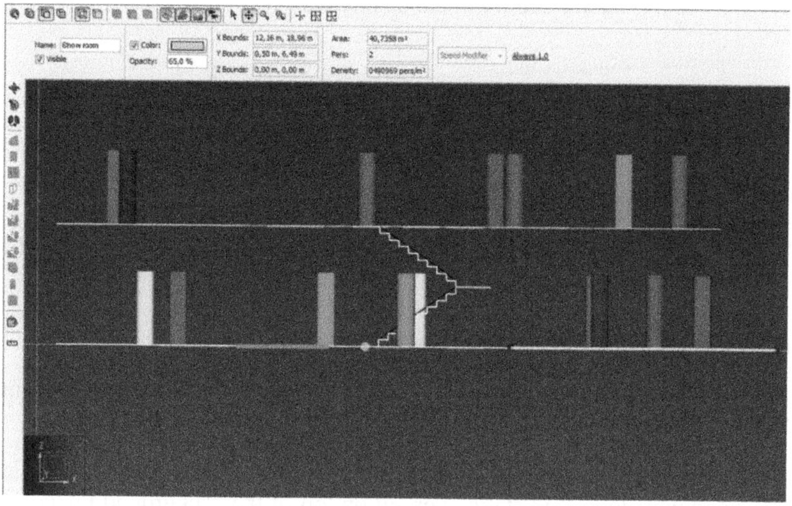

Figure 3.34. *Our model from the face-on view. We can see the stairs with its landing and the first and second floors. For a color version of this figure, see www.iste.co.uk/reveillac/modeling3.zip*

The SWITCH TO THE PERSPECTIVE VIEW icon, which appears when we are in the associated 3D ORBIT NAVIGATION mode, can also be used to pivot the view around the three axes.

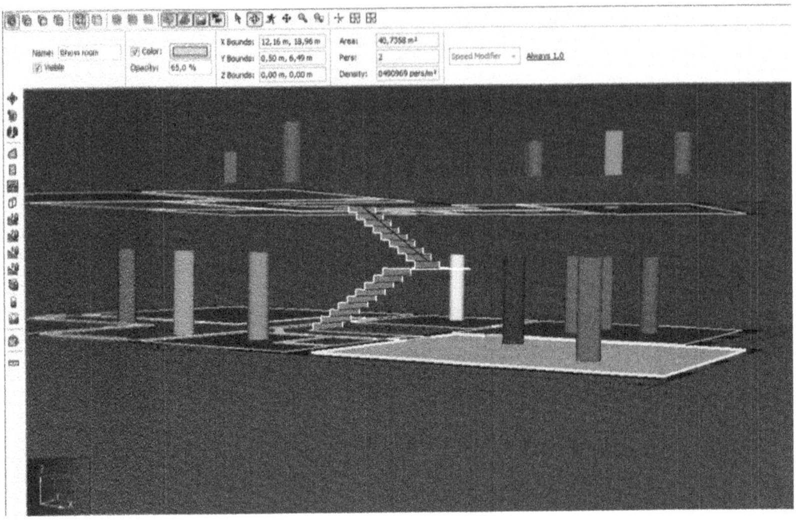

Figure 3.35. *Our model from a perspective view positioned using the orbit navigation tool. For a color version of this figure, see www.iste.co.uk/reveillac/modeling3.zip*

3.3.10. *Simulation and results for evacuation of the whole building*

The simulation can now be run as in section 3.3.5. The compilation/generation phase should complete without problem and take you to the simulation window.

The sequence should last around 30 s, with the occupants on the second floor leaving last, as expected.

3.3.10.1. *Visualization modes*

For optimized visualization, the visualization mode can be changed from the SCENE menu:

– SHOW NAV MESH OVER IMPORTED GEOM: in our case, we should see each of our rooms, doors and stairs overlaid on top of the background images made from the imported plans;

– SHOW IMPORTED GEOM OVER NAV MESH: in our case, we should see the doors and stairs in the form of inclined planes;

– SHOW NAV MESH ONLY: only see the components that we have drawn, such as the rectangular rooms, polygonal rooms and doors;

– SHOW IMPORTED GEOM ONLY: for our model, we only have the plans forming the background images for each level from the blueprints of the first and second floors;

– LAYOUT FLOORS VERTICALLY: place the floors vertically, as in real-life;

– LAYOUT FLOORS HORIZONTALLY: place the layers corresponding to the floors horizontally, side-by-side.

3.3.10.2. *Occupant characteristics and appearance*

The occupants in a model exhibit several characteristics which are adjustable. This mostly concerns their physical appearance, moving speed, 3D avatar, etc., which are gathered into a profile.

Figure 3.36 shows the list of the occupants, taken from the navigation tree, with the names given to them in the model.

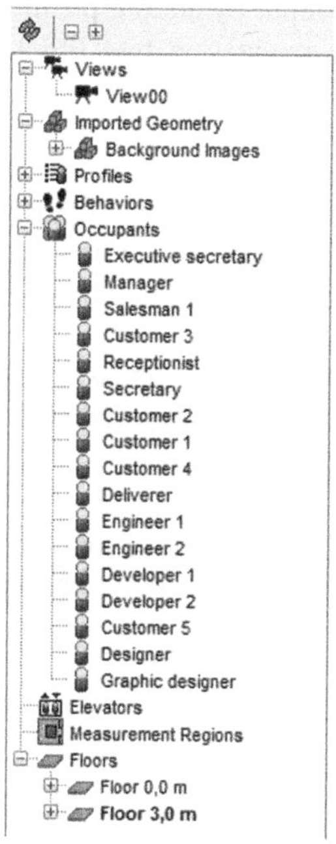

Figure 3.36. *The list of occupants in the model*

For better visualization of the different occupants, standard profiles have been defined (under PROFILES): male client, female client, male employee and female employee.

Name	3D Model	Color	Speed	Shoulder width	Description
Male customer	BMan0002 or BMan0003	Blue 0000FF	1.2 m/s	52 cm	Customer or people external to the company
Female customer	BWom0001 or BWom0002	Red FF0000	1.05 m/s	45 cm	Customer or people external to the company
Male employee	BMan0001	Cyan 00FFFF	1.2 m/s	52 cm	Company employee
Female employee	WWom0010	Green 00FF00	1.05 m/s	45 cm	Company employee

Table 3.1. *Characteristics for each of the occupant profiles*

All of these parameters are input by adding profiles into our model. Scroll down the PROFILES branch on the navigation tree (click on +), then double-click on DEFAULT.

COMMENT 3.11.– It is also possible to add a profile by selecting PROFILES in the navigation tree, then right-clicking and selecting ADD A PROFILE. In the dialog box that opens, simply input the name of the new profile in the NAME field. You will see that a profile can also be based upon a pre-existing one.

In the EDIT PROFILES window that opens, click on NEW, input the name of the new profile and then fill in the required parameter fields:

– NAME: name of the profile.

– 3D MODEL: click on EDIT… then, in the 3D MODEL dialog box, click on the desired 3D model(s). Unselected models will remain grayed out. Click on OK to confirm your choice.

– COLOR: in the COLOR dialog box, input the COLOR CODE and click on OK.

– DESCRIPTION: description of the profile (optional).

– SPEED: under the CHARACTERISTICS tab, the speed in m/s (units not required).

– SHOULDER WIDTH: under the CHARACTERISTICS tab, the shoulder width in cm (units not required).

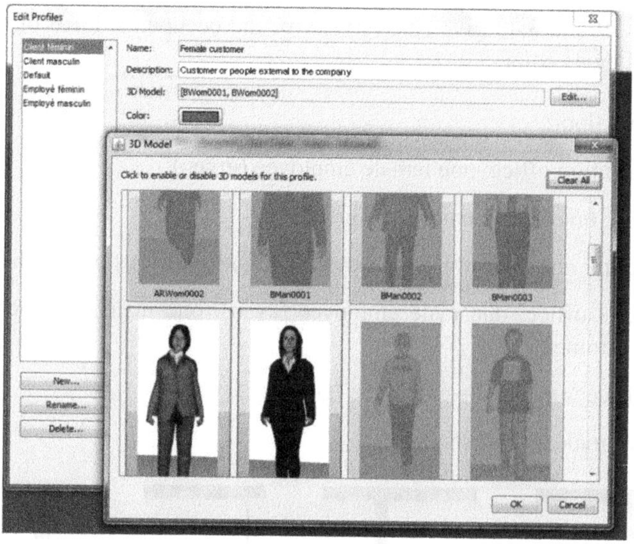

Figure 3.37. *The EDIT PROFILES window and its 3D MODEL dialog box*

COMMENT 3.12.–

– The speed can be set to CONSTANT or follow a statistical distribution (normal, logarithmic, etc.)

– Several other parameters exist that can be edited in the profile. These are largely beyond the scope of this book, and will only be mentioned if they do not take their default values.

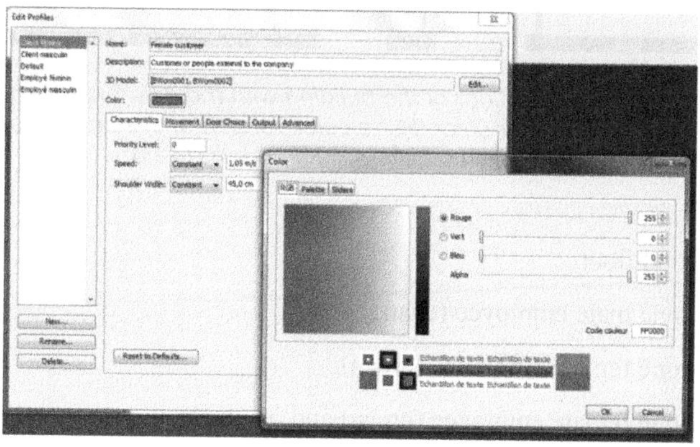

Figure 3.38. *The EDIT PROFILES window and its COLOR dialog box. For a color version of this figure, see www.iste.co.uk/reveillac/modeling3.zip*

Figures 3.39 and 3.40 give an overview of the occupants. Each of them can be distinguished by its color.

On the first floor:

– administration office: one female employee (green);

– reception: one female employee (green);

– show room: one male employee (cyan) and one male customer (blue);

– manager's office: one male employee (cyan), one male customer (blue) and one female customer (red);

– PA's office: one female employee (green);

– hall: two male customers (blue).

Figure 3.39. *The first floor of the building with the different occupants, identifiable by their color. For a color version of this figure, see www.iste.co.uk/reveillac/modeling3.zip*

On the second floor:

– office 1: one male employee (cyan);

– office 3: one female employee (green);

– office 4: one female employee (green) and one male customer (cyan);

– break room: one female employee (green) and one male employee (cyan);

– landing: one male employee (cyan).

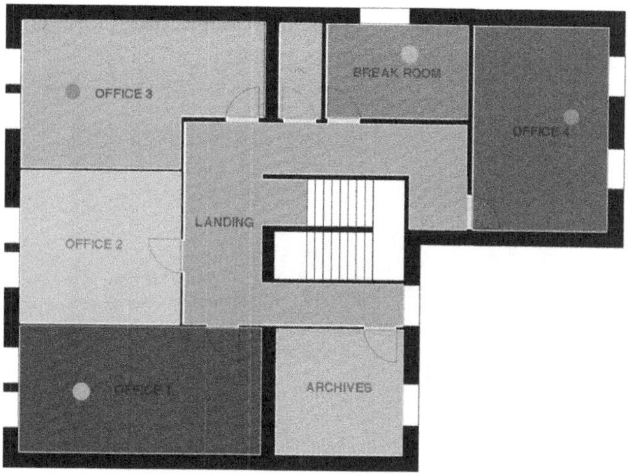

Figure 3.40. *The second floor of the building and its different occupants, identifiable by their color. For a color version of this figure, see www.iste.co.uk/reveillac/modeling3.zip*

3.3.10.3. *New simulation*

Rerunning the simulation with these new occupant parameters taken into account, the evacuation will take slightly longer, at around 32 s.

COMMENT 3.13.– During the simulation, it is still possible to use all of the visualization controls on the toolbar in the simulation window: TOP VIEW, FRONT VIEW, SHOW OUTLINES, 3D ORBIT NAVIGATION, PAN VIEW, 1ST PERSON ROAM MODE, etc.

3.3.10.4. *Visualizing the scene from an occupant's point of view*

During the simulation, visualization takes place from a third person perspective or objective viewpoint, i.e. the user sees the simulation run as a spectator watching the scene from afar.

A first person view also exists, which is also known as the subjective view. In this case, the user sees the simulation through the eyes of a selected occupant or from a camera he is carrying.

In the Pathfinder visualization window, you can watch the simulation from the first person perspective of an occupant.

Click on a person (they should turn yellow), then click on the SHOW CAMERA FEED FROM SELECTED OBJECT symbol (see Figure 3.20), which resembles a camera with a small green triangle. The triangle should turn into a green square and your view should move to just behind the occupant.

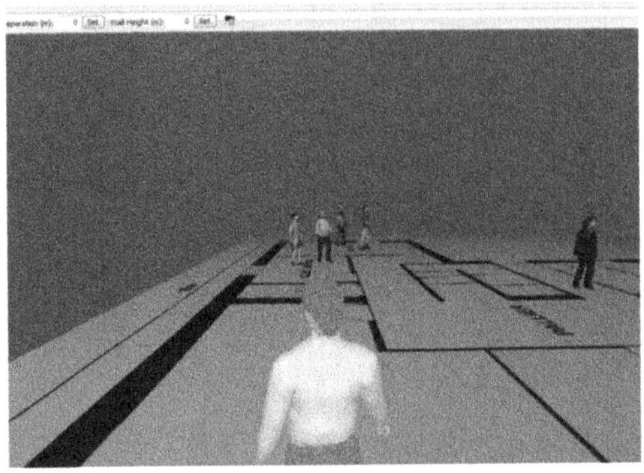

Figure 3.41. *A first person view from the point of view of an occupant. In the toolbar above, you can see the SHOW CAMERA FEED FROM SELECTED OBJECT symbol with a small green square*

If you run the simulation, you will see that you follow the occupant and you will see the evacuation unfold as if you were in the building looking through the different rooms.

3.3.10.5. *Modifying and creating new behavior*

Combined with their profile, the occupants also exhibit a behavior that influences how they evacuate the building.

These behaviors can, for example, determine specific choices taken by the occupant, such as taking the stairs rather than an elevator or *vice-versa*, or moving toward a particular exit, etc.

Our model contains two possible exits, the service entrance ('Service door') or the main door ('Main entrance door'). We will assign several occupants with a behavior forcing them to evacuate through the service entrance.

Right-click on BEHAVIORS in the navigation tree and select ADD A BEHAVIOR…

In the dialog box that opens, input 'Towards service entrance' in the NAME field and click on OK to confirm the creation.

The new behavior should appear under the BEHAVIORS branch in the navigation tree.

Select it and click on the ANY link found next to the EXITS field on the property panel. An EXITS dialog box should appear (Figure 3.42).

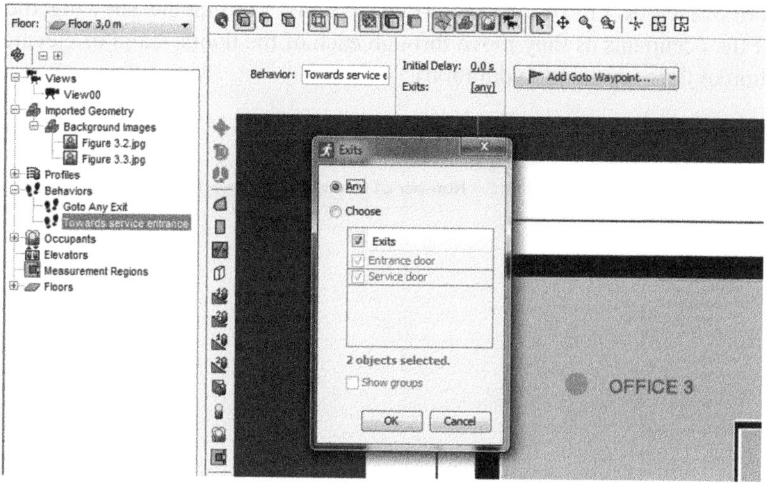

Figure 3.42. *The navigation tree with our 'Towards service entrance' behavior and the EXITS dialog box for modifying the behavior of the employees*

Check the CHOOSE radio button, then uncheck ENTRANCE DOOR and confirm by clicking on OK.

Next, using the SELECT/EDIT OBJECTS tool (see Figure 3.4) and the CTRL key, select the four employees on the second floor, then scroll down the BEHAVIOR list from the property panel and select TOWARDS SERVICE ENTRANCE instead of GOTO ANY EXIT.

If we rerun the simulation, after generating and compiling the scene, we can see that the employees on the second floor evacuate the building by the service entrance.

The evacuation lasts slightly longer, reaching around 35 s.

3.3.10.6. *Viewing and summarizing results*

During the simulation, Pathfinder stores a range of results that can be accessed later.

After closing the simulation window, you can obtain two types of graphics by going into the RESULTS menu:

– VIEW ROOM USAGE (Figure 3.43): 2D graphic showing the number of occupants in the different rooms as a function of time (evacuation duration);

– VIEW DOOR FLOW RATES (Figure 3.44): 2D graphic showing the flow rates of the occupants as they move through each of the doors, stairs or elevators, as a function of time (evacuation duration).

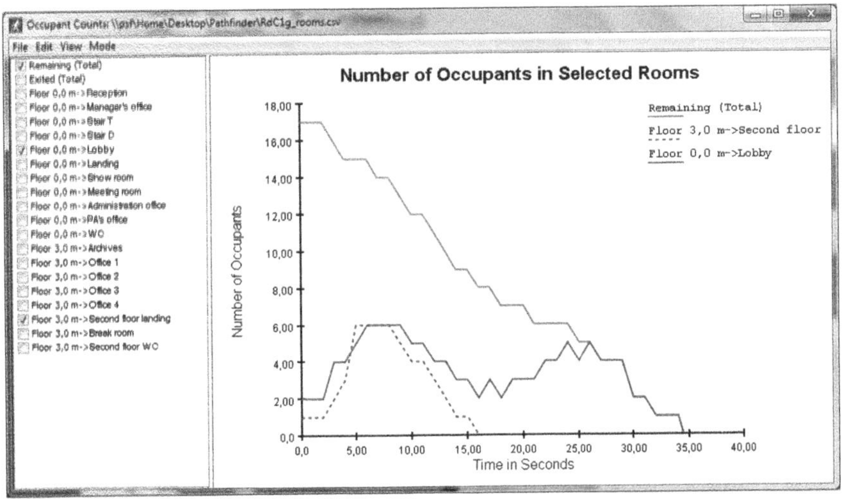

Figure 3.43. *The number of occupants on the second floor landing and the first floor lobby as a function of time. For a color version of this figure, see www.iste.co.uk/reveillac/modeling3.zip*

For each of these graphics, you can select the elements of interest by checking them on a tree menu.

There is also a file available that summarizes all of the information pertaining to the simulation: the number of occupants, journey time, duration of the evacuation, etc.

This file is called the SUMMARY FILE and it is accessible from the RESULTS menu, by selecting SHOW SUMMARY FILE. It can also be saved by selecting FILE and then SAVE in the report window that opens (Figure 3.45). The file is saved in .TXT format, so that it can be easily reopened by a classical text editor or word processor.

Figure 3.46 shows an example of the summary produced by the simulation of our model, opened and formatted in Microsoft Word (tabulation, spacing, etc.).

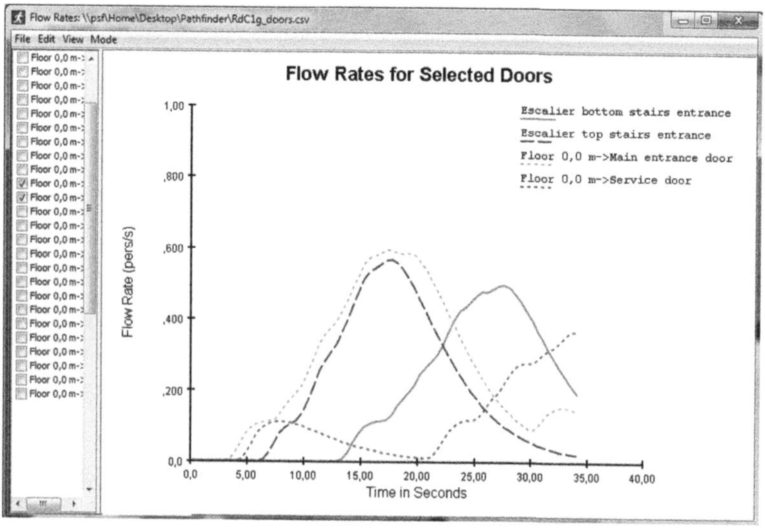

Figure 3.44. *The flow rates for the upper and lower access points to the stairs, and for the exit doors. For a color version of this figure, see www.iste.co.uk/reveillac/modeling3.zip*

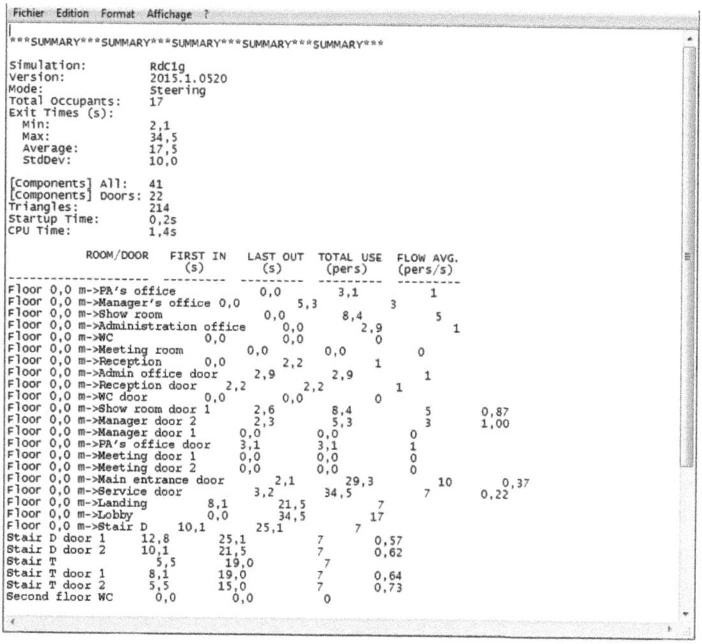

Fichier Edition Format Affichage ?

SUMMARYSUMMARY***SUMMARY***SUMMARY***SUMMARY***

Simulation:	RdC1g
Version:	2015.1.0520
Mode:	Steering
Total Occupants:	17
Exit Times (s):	
Min:	2,1
Max:	34,5
Average:	17,5
StdDev:	10,0
[Components] All:	41
[Components] Doors:	22
Triangles:	214
Startup Time:	0,2s
CPU Time:	1,4s

ROOM/DOOR	FIRST IN (s)	LAST OUT (s)	TOTAL USE (pers)	FLOW AVG. (pers/s)
Floor 0,0 m->PA's office	0,0	3,1	1	
Floor 0,0 m->Manager's office	0,0	5,3	3	
Floor 0,0 m->Show room	0,0	8,4	5	
Floor 0,0 m->Administration office	0,0	2,9	1	
Floor 0,0 m->WC	0,0	0,0	0	
Floor 0,0 m->Meeting room	0,0	0,0	0	
Floor 0,0 m->Reception	0,0	2,2	1	
Floor 0,0 m->Admin office door	2,9	2,9	1	
Floor 0,0 m->Reception door	2,2	2,2	1	
Floor 0,0 m->WC door	0,0	0,0	0	
Floor 0,0 m->Show room door 1	2,6	8,4	5	0,87
Floor 0,0 m->Manager door 2	2,3	5,3	3	1,00
Floor 0,0 m->Manager door 1	0,0	0,0	0	
Floor 0,0 m->PA's office door	3,1	3,1	1	
Floor 0,0 m->Meeting door 1	0,0	0,0	0	
Floor 0,0 m->Meeting door 2	0,0	0,0	0	
Floor 0,0 m->Main entrance door	2,1	29,3	10	0,37
Floor 0,0 m->Service door	3,2	34,5	7	0,22
Floor 0,0 m->Landing	8,1	21,5	7	
Floor 0,0 m->Lobby	0,0	34,5	17	
Floor 0,0 m->Stair D	10,1	25,1	7	
Stair D door 1	12,8	25,1	7	0,57
Stair D door 2	10,1	21,5	7	0,62
Stair T	5,5	19,0	7	
Stair T door 1	8,1	19,0	7	0,64
Stair T door 2	5,5	15,0	7	0,73
Second floor WC	0,0	0,0	0	

Figure 3.45. *The visualization window of the file summarizing the simulation data*

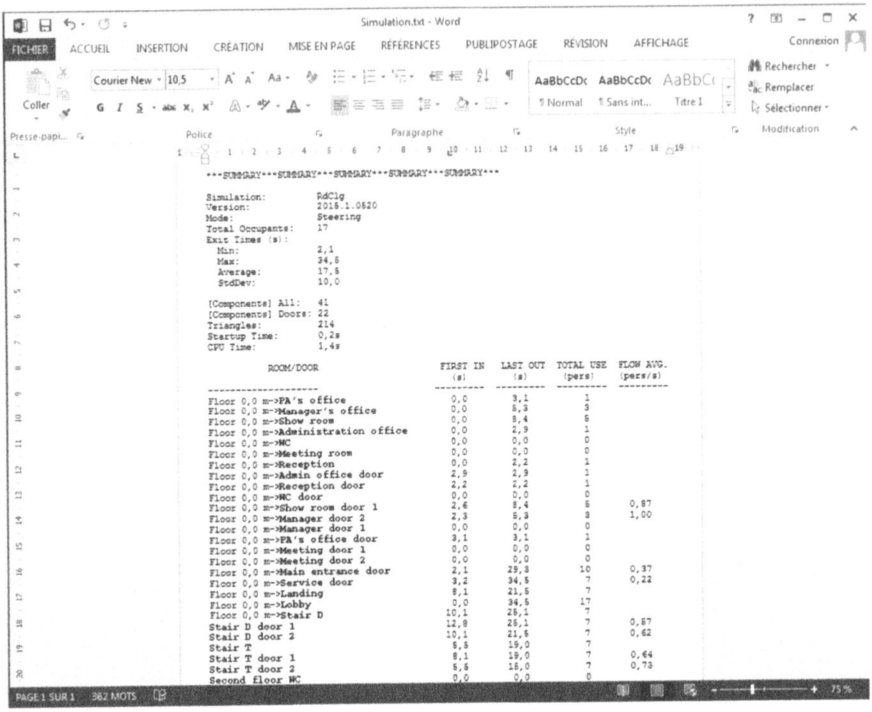

SUMMARYSUMMARY***SUMMARY***SUMMARY***SUMMARY***

Simulation: RdC1g
Version: 2015.1.0520
Mode: Steering
Total Occupants: 17
Exit Times (s):
Min: 2,1
Max: 34,5
Average: 17,5
StdDev: 10,0

[Components] All: 41
[Components] Doors: 22
Triangles: 214
Startup Time: 0,2s
CPU Time: 1,4s

ROOM/DOOR	FIRST IN (s)	LAST OUT (s)	TOTAL USE (pers)	FLOW AVG. (pers/s)
Floor 0,0 m->PA's office	0,0	3,1	1	
Floor 0,0 m->Manager's office	0,0	5,3	3	
Floor 0,0 m->Show room	0,0	8,4	5	
Floor 0,0 m->Administration office	0,0	2,9	1	
Floor 0,0 m->WC	0,0	0,0	0	
Floor 0,0 m->Meeting room	0,0	0,0	0	
Floor 0,0 m->Reception	0,0	2,2	1	
Floor 0,0 m->Admin office door	2,9	2,9	1	
Floor 0,0 m->Reception door	2,2	2,2	1	
Floor 0,0 m->WC door	0,0	0,0	0	
Floor 0,0 m->Show room door 1	2,6	8,4	5	0,87
Floor 0,0 m->Manager door 2	2,3	5,3	3	1,00
Floor 0,0 m->Manager door 1	0,0	0,0	0	
Floor 0,0 m->PA's office door	3,1	3,1	1	
Floor 0,0 m->Meeting door 1	0,0	0,0	0	
Floor 0,0 m->Meeting door 2	0,0	0,0	0	
Floor 0,0 m->Main entrance door	2,1	29,3	10	0,37
Floor 0,0 m->Service door	3,2	34,5	7	0,22
Floor 0,0 m->Landing	8,1	21,5	7	
Floor 0,0 m->Lobby	0,0	34,5	17	
Floor 0,0 m->Stair D	10,1	25,1	7	
Stair D door 1	12,8	25,1	7	0,57
Stair D door 2	10,1	21,5	7	0,62
Stair T	5,5	19,0	7	
Stair T door 1	8,1	19,0	7	0,64
Stair T door 2	5,5	15,0	7	0,73
Second floor WC	0,0	0,0	0	

Figure 3.46. *The summary file from our model, imported and formatted in Microsoft Word 2013*

On the right-hand side of the simulation window, there is a tree menu showing the cameras and the available viewpoints (VIEWPOINTS), as well as an OCCUPANT CONTOURS branch, giving a 'heat map' around each occupant.

This heat map is visible during the simulation in the form of colored zones appearing around each occupant. A legend on the right-hand side of the model shows the relationship between the relevant values and colors.

The following contours are available:

– DENSITY: density of occupants (the number of occupants sharing a surface area), graded from 0 to 3;

– LEVEL OF SERVICE (Queueing): index of room occupation, graded from A to F;

– LEVEL OF SERVICE (Stairway): index of stair occupation, graded from A to F;

– LEVEL OF SERVICE (Walkway): index of floor occupation during movement, graded from A to F;

– NORMALIZED SPEED: speed of the occupants, graded from 0 to 1;

– SPEED: speed of the occupants in m/s, graded from 0 to n (n: maximum speed in m/s);

– TIME TO EXIT: time (in seconds) to evacuate or leave the building;

– USAGE (Accumulated): cumulative occupant density in the different rooms, as a function of time passed (during and after the simulation);

– USAGE (Instantaneous): instantaneous occupant density in the different rooms, as a function of time passed (during the simulation).

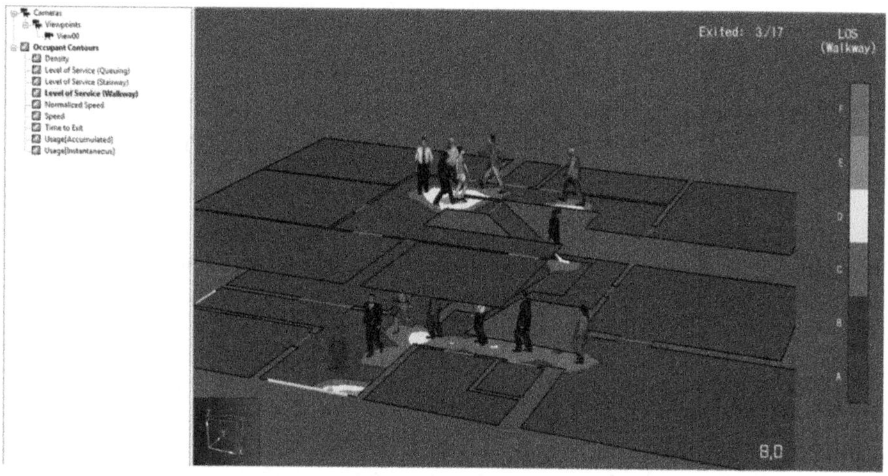

Figure 3.47. *Our model in simulation mode. Here, the walkway level of service index is being displayed, ranging from A (violet) to E (orange, at the top of the stairs). On the left, the simulation tree is visible, and on the right so too is the contour legend. For a color version of this figure, see www.iste.co.uk/reveillac/modeling3.zip*

COMMENT 3.14.– To go from the creation/modeling window to the simulation window, it is not necessary to repeat the generation/compilation by clicking on the RUN SIMULATION icon every time. Simply clicking on VIEW 3D RESULTS (Figure 3.4) in the main toolbar is enough.

3.4. Extensions

The Pathfinder software offers numerous other options, such as simulating in SFPE mode, managing groups of occupants and managing the elevators.

The following sections cover some of the uses and motivations for these options, which our model has not made use of.

3.4.1. *Moving to SFPE mode*

In the example we have just looked at, we used the default simulation mode known as 'Steering mode'.

This mode is based upon a computer simulation program developed by Craig W. Reynolds in 1986, called 'Boids'. His aim was to show that complex behavior, such as the coordinated flight of a flock of birds, arises from a collection of local, very simple rules. In the bird example, each bird's behavior is determined by the behavior of its immediate neighbors.

These theories are applied today, notably for simulating crowd behavior, whether for video games, blockbuster films featuring large numbers of people, or indeed the evacuation of a building.

Reynolds also presented a behavioral study in an article dating from 1999 called 'Steering behaviors for autonomous characters'[2], which has become a reference text for specialists in building evacuation. This was improved[3] and developed in 2006 by Heni Ben Amor, Ian Murray and Oliver Obst.

Another often-used mode exists, which refers to the SFPE (Society of Fire Protection Engineers) manual by Nelson and Mowrer, dating from 2002. In this mode, the occupants move at speeds determined by the occupant density in each room, and the flow through doors is a function of their widths.

If you go to the SIMULATION menu, and then SIMULATION PARAMETERS, a dialog box containing five tabs appears. Go to the BEHAVIOR tab and select SFPE from the BEHAVIOR MODE drop-down list. Confirm your choice by clicking on OK. The model will now be configured for SFPE mode, with all of its default parameters.

---

2 Downloadable from: http://www.red3d.com/cwr/papers/1999/gdc99steer.pdf or http://www.red3d.com/cwr/steer/gdc99/.

3 Downloadable from: http://www.uni-koblenz-landau.de/de/koblenz/fb4/ifi/AGKI/ Publications/2006/shortsteering.pdf.

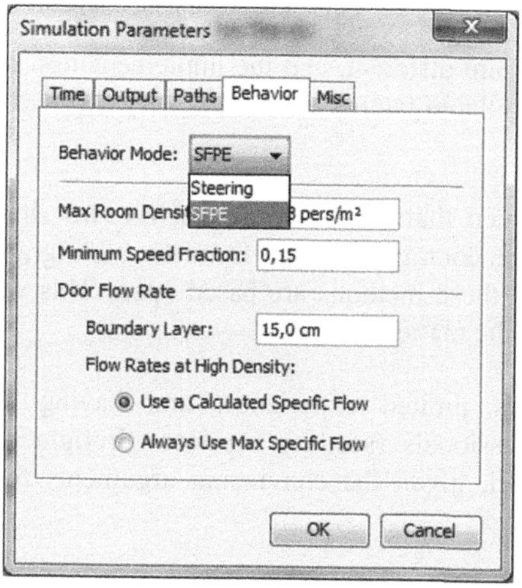

Figure 3.48. *The dialog box for changing the mode*

COMMENT 3.15.– To confirm the change of mode, do not forget to rerun the model generation/compilation via RUN SIMULATION.

The effects of changing modes can be seen in the simulation results shown in Figure 3.49.

```
***SUMMARY***SUMMARY***SUMMARY***SUMMARY***SUMMARY***

Simulation:            Rdc1h
Version:               2015.1.0520
Mode:                  Steering
Total Occupants:       17
Exit Times (s):
  Min:                 2,1
  Max:                 34,5
  Average:             17,5
  StdDev:              10,0

[Components] All:      41
[Components] Doors:    22
Triangles:             214
Startup Time:          0,2s
CPU Time:              1,3s
```

```
***SUMMARY***SUMMARY***SUMMARY***SUMMARY***SUMMARY***

Simulation:            Rdc1h
Version:               2015.1.0520
Mode:                  SFPE (Basic)
Total Occupants:       17
Exit Times (s):
  Min:                 1,5
  Max:                 30,5
  Average:             15,6
  StdDev:              9,1

[Components] All:      41
[Components] Doors:    22
Triangles:             214
Startup Time:          0,1s
CPU Time:              1,0s
```

Figure 3.49. *The top part of the two results summaries for the same model, in 'steering mode' (above) and 'SFPE' mode (below)*

This chapter cannot tell you which mode is the most appropriate. The calculation parameters are different, and the implementation of each mode depends upon so many uncontrollable factors that is it is very difficult to favor one over the other.

It must also be noted that while Pathfinder has two simulation modes, many others exist. You are encouraged to read from the bibliography at the end of this book to find out what these methods are based upon. This will allow you to form your own opinions on the matter.

From the author's limited experience, and having discussed the matter with evacuation professionals (security services, firefighters, etc.), no objective opinion can be offered, given the convincing arguments for all of the different modes used.

3.4.2. *Groups of occupants*

It is possible to define groups of occupants to be placed within a rectangular area marked out inside a room.

The benefit of this technique is the ability to assign the same behavior and profile to a group of occupants.

The occupant number and density relative to the chosen surface are configurable, as well as the occupants' positioning.

The ADD RECTANGULAR GROUP OF OCCUPANTS tool (see Figure 3.10) in the creation toolbar is used to create groups.

The property panel displays all of the available parameters for the group: NAME, COUNT, PLACEMENT, PROFILE, BEHAVIOR and DENSITY.

Figure 3.50 shows an example of two groups placed in the first floor meeting room in our model.

3.4.3. *Managing the elevators*

Pathfinder offers the option of easily modeling the multiple elevators often available in buildings, by attributing them with numerous parameters so that they can match real life as closely as possible.

This section does not give a step-by-step guide for the example like the previous ones, but more of an overview of the general construction approach and the main concepts for managing the elevators.

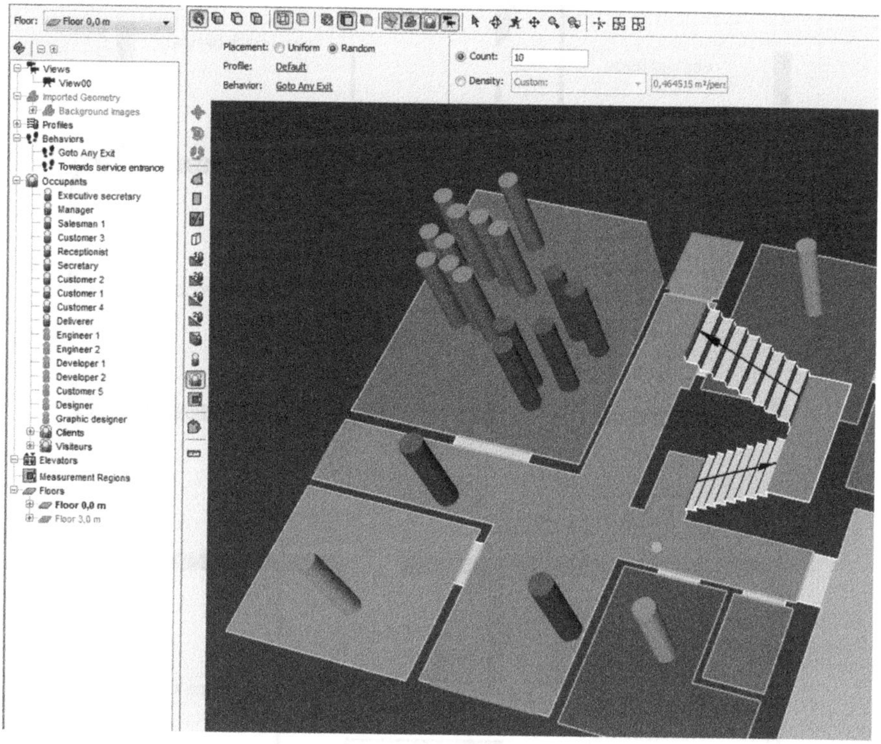

Figure 3.50. *The two groups, 'Clients' (orange) and 'Visitors' (fuchsia), in the first floor meeting room (top-left of the 3D model). Above, the property panel during the creation of a group. For a color version of this figure, see www.iste.co.uk/reveillac/modeling3.zip*

We could place an elevator in our example building as shown on the plans in Figure 3.51, in the lobby, opposite the entrance, to connect to the second floor landing. The meeting room and office #3 on the second floor would need to be modified in this case.

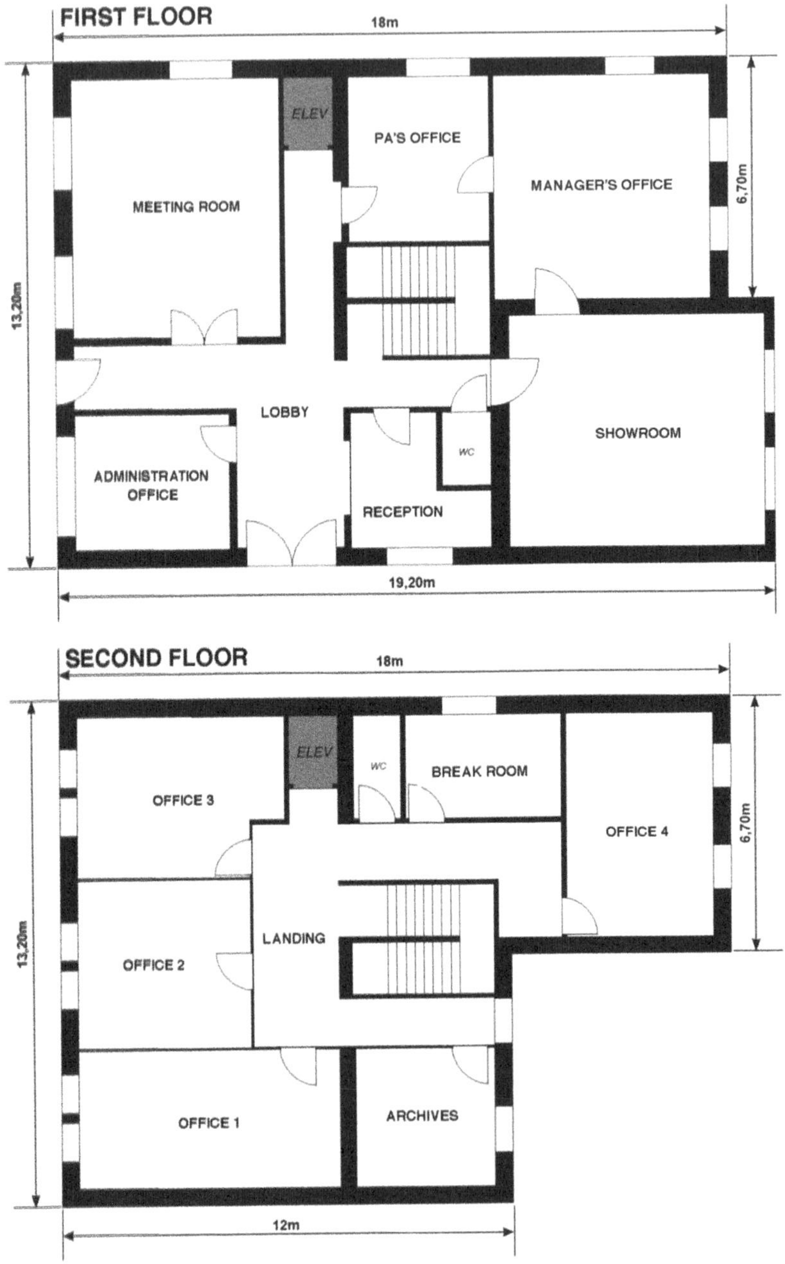

Figure 3.51. *The modified plans (first and second floors) for the installation of an elevator (in red). For a color version of this figure, see www.iste.co.uk/reveillac/modeling3.zip*

3.4.3.1. *Construction method for an elevator*

1) Switch to visualization mode: top-down view.

2) Go to the floor that the elevator will leave from.

3) Draw an outline representing the elevator shaft.

4) Add an access door.

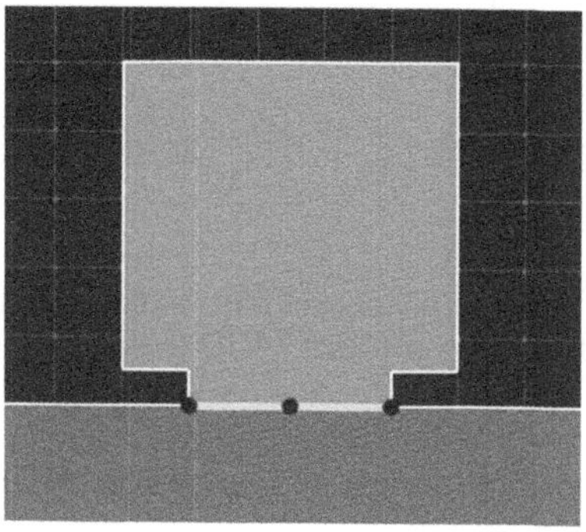

Figure 3.52. *An elevator shaft (two conjoined rectangles) and its access door*

5) Select the outline you have created and turn it into an elevator (right-click, CREATE ELEVATOR…).

6) Give the elevator a name and enter its different parameters (maximum number of occupants, movement bounds, acceleration, speed and opening/closing times).

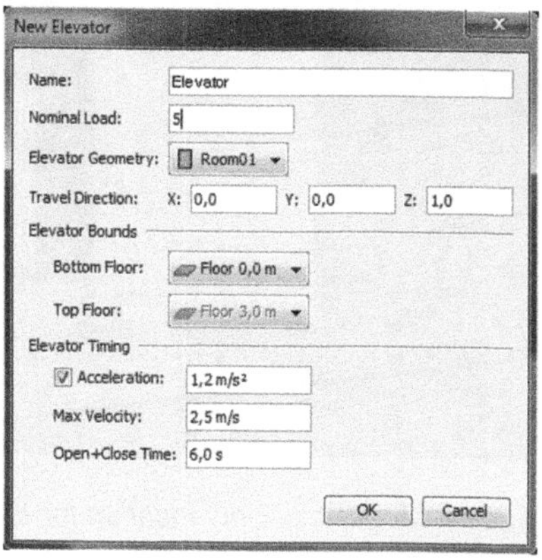

Figure 3.53. *The dialog box for creating and configuring the elevator*

7) Create a new behavior by associating it with the ADD GOTO ELEVATORS… action.

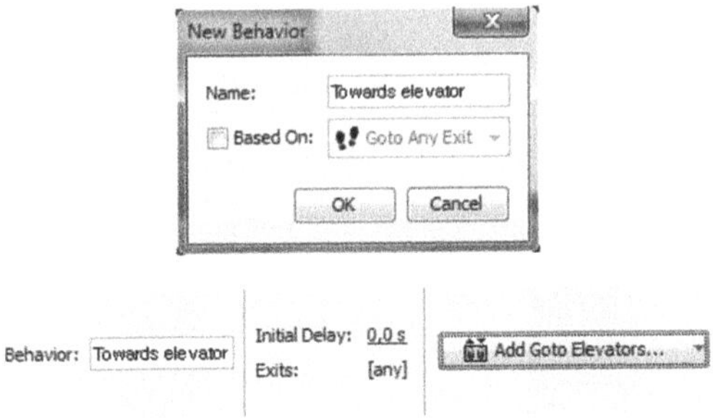

Figure 3.54. *New behavior and association with the action of movement toward the elevator (ADD GOTO ELEVATORS…)*

8) Select the occupants or groups of occupants and assign them the newly created behavior.

9) Select the priority entrance for the elevator (intermediate level, first floor and top floor).

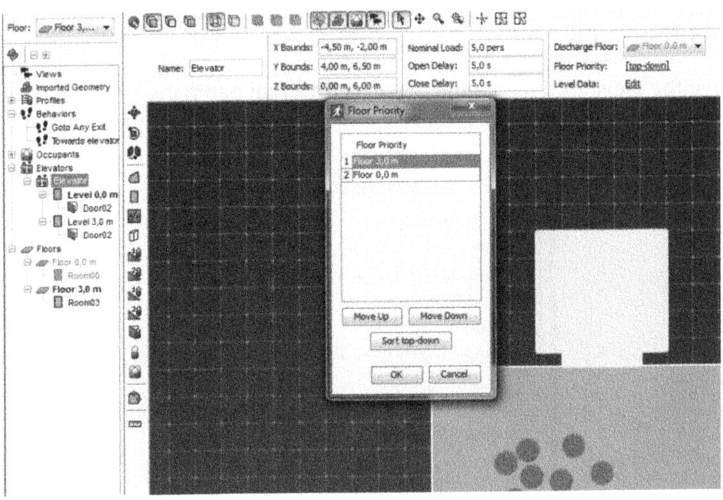

Figure 3.55. *Selecting the priority location for the elevator*

10) Launch the simulation generation/compilation.

3.4.3.2. *Available options for elevator usage*

The presence of one or more elevators in a building leads to new specifications that can, in most cases, be addressed using the numerous parameters and options offered in the Pathfinder software.

The elevators are generally created once the rest of the model is already in place.

If you draw an elevator over existing geometry (rooms), the software will automatically create a hole through this on all floors to model the elevator shaft. In this case, the doors, ramps, stairs and rooms that are in the way will be deleted once confirmed *via* a dialog box.

For an occupant or a group of occupants to use an elevator, you must specify this in their behavior.

The parameters attributed to an elevator are the same that have already been presented in point 6 of section 3.4.3.1, in detail:

– NAME: elevator name;

– NOMINAL LOAD: maximum number of occupants (estimated) that the elevator can contain;

– ELEVATOR GEOMETRY: geometric shape of the elevator floor;

– TRAVEL DIRECTION: vector defining the elevator motion. By default, this is a vector normal to the ground (negative values are permitted);

– ELEVATOR BOUNDS: defines the values of each of the floors that the elevator can reach in its motion;

– BOTTOM FLOOR: the lowest floor taken into account during elevator motion;

– TOP FLOOR: the highest floor taken into account during elevator motion;

– ELEVATOR TIMING: collection of basic time parameters taken into account by the model for loading and then dropping off the occupants between two floors;

– ACCELERATION: acceleration of the elevator motion in meters per second squared (m/s^2). This value is optional (applicable if the box is checked);

– MAX VELOCITY: maximum speed of the elevator motion in meters per second (m/s);

– OPEN+CLOSE TIME: sum of the opening and closing times for the elevator door (each of these two values equals half of the time specified) in seconds (s).

After being created, the elevator appears in the navigation tree, in the ELEVATORS branch.

If you select the elevator, you can see all of its parameters in the property panel.

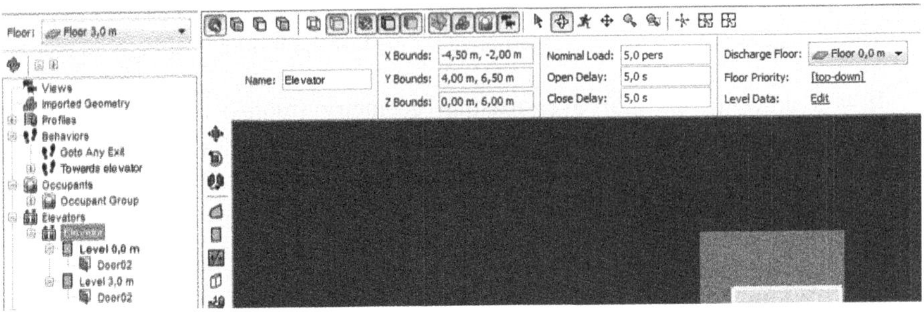

Figure 3.56. *The selected elevator and its associated parameters*

In the navigation tree for the elevator, you can introduce branches assigned to each of the floors. These branches can be reconfigured by selecting them and inputting the appropriate attributes in the different fields available in the property panel.

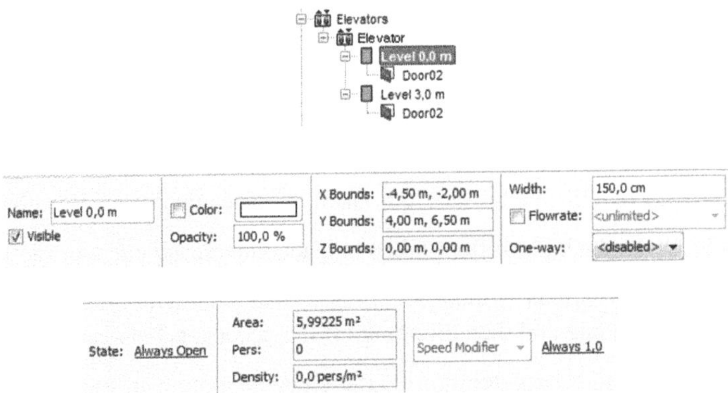

Figure 3.57. *The elevator navigation tree and its property panel*

3.4.4. *Creating viewpoints*

The model created in Pathfinder can be viewed from different angles using the visualization tools.

For a complex model, it can be useful to be able to place one or more cameras in strategic locations to obtain viewpoints showing specific parts of the model (exits, corridors, elevator entrances, overall view, etc.).

The VIEWS branch on the navigation tree is used for these purposes.

After moving to the desired point in the model, you can create a new viewpoint by right-clicking on the VIEWS branch in the navigation tree and selecting NEW VIEW.

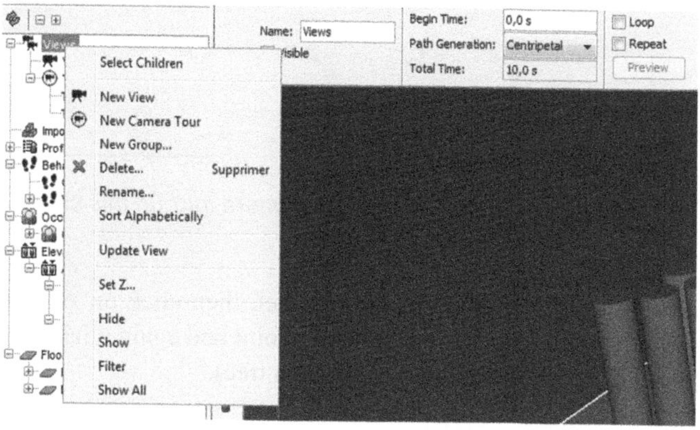

Figure 3.58. *The creation menu for a new viewpoint (NEW VIEW)*

Once the new viewpoint has been created, you can refine its location parameters in the property panel and give it a name if desired.

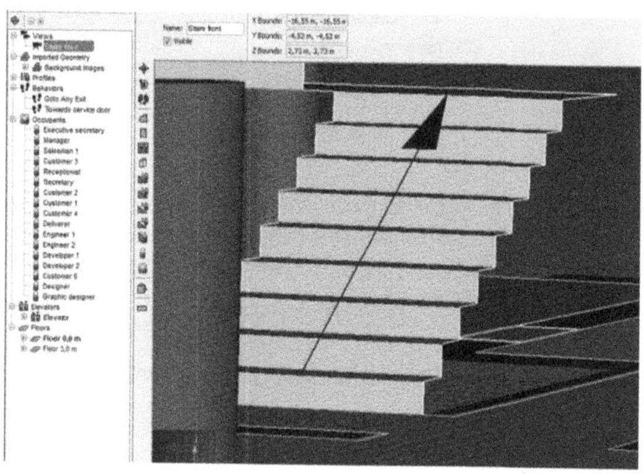

Figure 3.59. *A new viewpoint, called 'Stairs front'*

3.4.5. *Creating camera tours*

As well as creating viewpoints, it is also possible to define animated tours around your model. In this case, a camera path is established between viewpoints.

To create a camera tour, right-click on the VIEWS branch from the navigation tree, then select NEW CAMERA TOUR.

You should see a new tour appear in the property panel, which will be called TOUR00 by default, but can be renamed.

Figure 3.60. *The property panel for a new camera tour (in this case: 'Tour00')*

Move to the starting viewpoint in your model, then click on ADD VIEW from the property panel. Then, move to the end viewpoint and again click on ADD VIEW (make sure your tour is selected in the navigation tree).

Two new branches appear under your camera tour in the navigation tree. By default, the added viewpoints are called 'TOUR VIEW 1' and 'TOUR VIEW 2'. These can be renamed, or even reconfigured, using the UPDATE VIEW button in the property panel.

Figure 3.61. *The VIEWS tree which contains the viewpoints ('Stairs front'), the camera tours ('Tour00') and the viewpoints on the camera tour ('Tour View 1' and 'Tour View 2')*

Having selected your tour from the navigation tree, you can preview it by clicking on the PREVIEW button.

Several fields are available in the property panel, allowing you to adjust all of the parameters of the tour.

For a tour:

– BEGIN TIME: delay before the tour starts. Only occurs once if the tour is set to repetition mode;

– PATH GENERATION: shape of the camera's path between the viewpoints (UNIFORM, CHORDAL, CENTRIPETAL);

– TOTAL TIME: total duration of the tour, excluding repetitions;

– LOOP: the tour runs in loops (forward and then backward);

– REPEAT: repeats the tour indefinitely (starting from the starting viewpoint each time);

– PREVIEW: launches the tour preview;

– STOP PREVIEW: stops the tour preview;

– ADD VIEW: adds a new viewpoint to the pre-existing list;

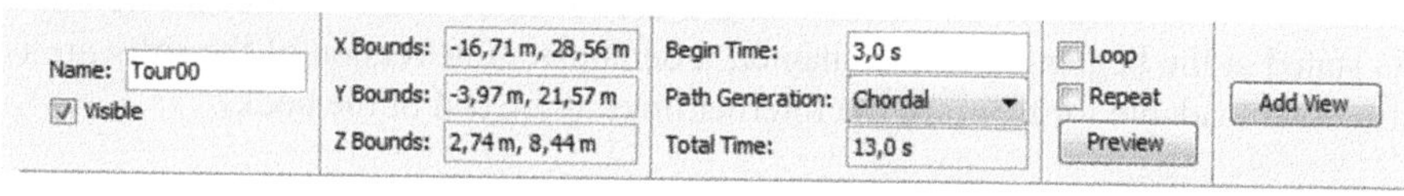

Figure 3.62. *The property panel associated with a camera tour branch (in this case: 'Tour00')*

For a viewpoint:

– TRANSITION TIME: time taken by the camera to reach this viewpoint;

– WAIT TIME: time the camera rests at this viewpoint;

– UPDATE VIEW: updates the camera position for this viewpoint (if the camera position has been changed);

– INSERT VIEW: inserts a new viewpoint after the currently selected one;

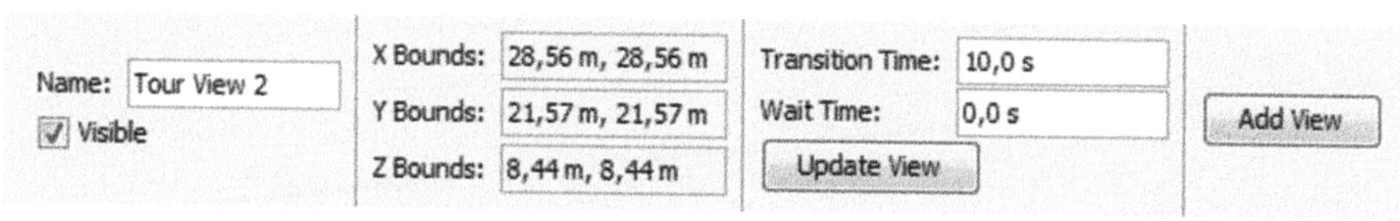

Figure 3.63. *The property panel associated with a viewpoint branch in a camera tour (in this case: 'Tour view 2')*

The camera tour playback panel is displayed by double-clicking on the branch containing the camera tour in the navigation tree.

Figure 3.64. *The camera tour playback panel with six functions: pause, stop, skip backward, rewind, fast-forward and skip ahead*

3.4.6. *Further possibilities*

Thunderhead Engineering's Pathfinder software is capable of working with much more sophisticated simulations than the examples that have been presented here.

Numerous exercises and examples are available on the publisher's Website as well as various documentation, including a user's guide and a technical reference. As stated at the beginning of the chapter, a demonstration version of the software is also downloadable for free (see the Internet links at the end of the book).

You can thus go further and explore all of the software functionality in more detail.

4

3D Flows, Distribution and Warehousing

4.1. Product distribution

In today's world of ubiquitous product distribution, the simulation of logistics flows is now realizing its full potential. Many companies are built around platforms for stocking all sorts of products.

With the advent of e-commerce, mail order shopping has exploded and created new distribution heavyweights (Amazon, Alibaba, AliExpress, Ebay, Priceminister, Rakuten, Cdiscount, etc.) that require robust logistics to provide optimum customer satisfaction. Delivery times must be cut to as short as possible for ever-cheaper costs, even as the number of products available grows greater and greater.

Warehousing and storage have become key operations and the focus of keen commercial interest, with sophisticated techniques being implemented and 3D flow simulation taking on critical importance.

In this chapter, we will focus on product distribution in a warehouse featuring classic processes such as unloading, loading, removing from pallets, packaging, retrieval, storage, dispatching, etc.

The FlexSim software has been chosen to support the example to be addressed in this chapter. This is due to its great flexibility and usability, which make it a simple and very powerful tool.

4.2. The FlexSim software

FlexSim is a simulation software program that has existed since February 2003. It integrates a 3D modeling environment with multiple functionalities.

Fundamentally, it is a tool that works with discrete events. A specific version for the medical market, called FlexSim Healthcare, has been available since 2009.

The FlexSim software can import its own 3D objects to give personalized simulation models. Furthermore, powerful analysis tools are available as well as the ability to generate graphics and reports linked to statistical results.

The FlexSim Software Products company has its headquarters in the USA, at Orem, Utah, and has many offices around the world.

A trial version of the software can be downloaded from the following address[1]: https://www.flexsim.com/fr/.

COMMENT 4.1.–

– The FlexSim France office is gratefully acknowledged, for providing the education edition of version 4.5, and later 4.7, of the software, free of charge for testing and developing the examples featured in this book.

– The version used below is version 4.7. If you are using a different version (older or newer), there may be some differences.

4.3. Basic concepts of the FlexSim software

As mentioned in the above introduction, the FlexSim software deals with discrete events, meaning that it assumes that events happen at specific times in a defined order, and never exactly simultaneously.

4.3.1. *General appearance of FlexSim*

Upon opening FlexSim, a start screen appears with a menu of different options allowing you to: create a new model (NEW MODEL), open an existing model (OPEN MODEL), access a guide to getting started with FlexSim (GETTING STARTED), access the preferences (PREFERENCES), manage licenses (LICENSING) or open a user's guide (USER MANUAL).

After selecting NEW MODEL, a dialog box is displayed. Click on OK to confirm the default model units suggested. We will return to this later on in the chapter.

1 This link works at the time of writing. Over time, it may change or be removed, in which case search for "FlexSim" in your search engine.

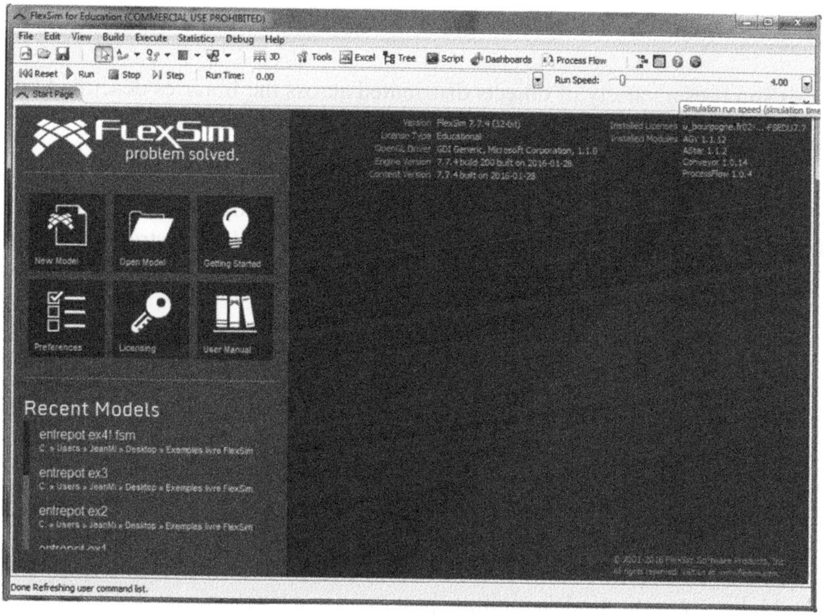

Figure 4.1. *The start screen that appears after launching FlexSim*

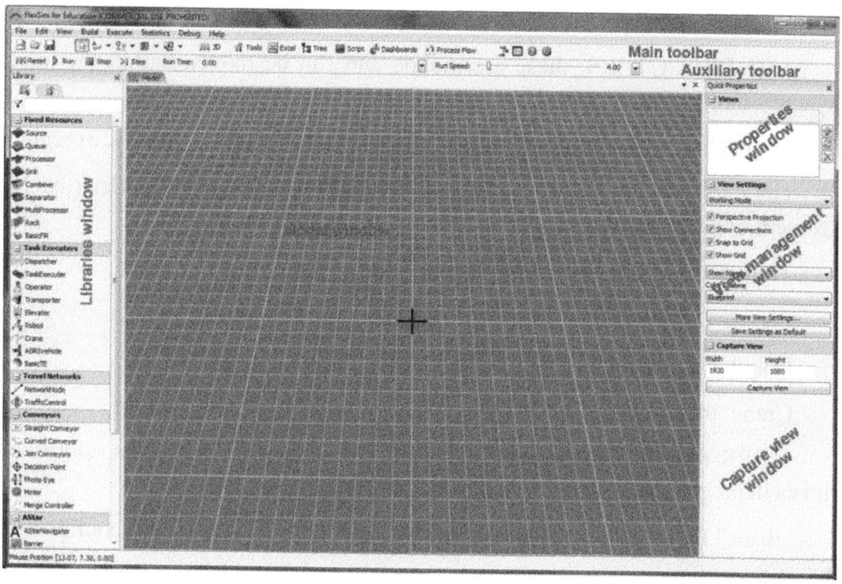

Figure 4.2. *The default main FlexSim display with its various windows and toolbars*

4.3.2. *Libraries*

FlexSim uses a variety of objects distributed across the libraries described below (DRAG-DROP LIBRARY). Only the most widely used are included here.

– FIXED RESOURCES: These are stationary objects in the model and their main functions are to create, store, assemble, divide, destroy, etc. They can push or pull flows and communicate with task executers and pass them instructions or tasks such as to transport, call or send:

- Source: generates items[2];
- Queue: models a queue or a place where items can wait;
- Processor: models a treatment to be applied to the items;
- Sink: destroys items;
- Combiner: combines items (packaging, palletizing, merging, batching, etc.);
- Separator: splits items (unpacking, depalletizing, etc.);
- Multiprocessor: models multiple processes in a single place;
- Rack: stores items (on the floor, on shelves, etc.);
- BasicFR: models a general fixed resource, with advanced configuration options.

– TASK EXECUTERS: these are mobile and can be moved within the model. They carry out the orders or tasks given to them by the fixed resources.

- Dispatcher: receives orders or tasks and delegates them to the task executers;
- Operator: operates and/or transports an item, carries out an order (machine adjustment, maintenance, etc.);
- Transporter: transports item(s);
- Elevator: transports item(s) vertically;
- Robot: carries out a repetitive, automated operation;
- Crane: transports items;
- ASRS (Automated Storage and Retrieval System): transports, stores and retrieves items placed on shelves or storage racks;
- BasicTE: models a general task executer, with advanced configuration options.

---

2 The concept of items is very general. An item can be a product, a cardboard box, a pallet, a container, a vehicle, a person, etc.

– TRAVEL NETWORKS:

- NetworkNode: defines network nodes. These nodes can be connected to form paths. A fixed resource or task executer can be connected to a network;

- TrafficControl: controls flow traffic around a nodal network.

– CONVEYORS[3]:

- Straight conveyor: transports items in a straight line;

- Curved conveyor: transports items in a curved line;

- Join conveyors: groups and joins conveyors together;

- Decision Point: used to construct a decision point for distributing items around a network of conveyors (directing an item to specific conveyors according to one or more criteria);

- Photo Eye: detects an item placed on a conveyor entailing four states, which can trigger different actions;

- Motor: activates or deactivates a conveyor according to certain conditions;

- Merge Controller: merges transport routes from multiple conveyors.

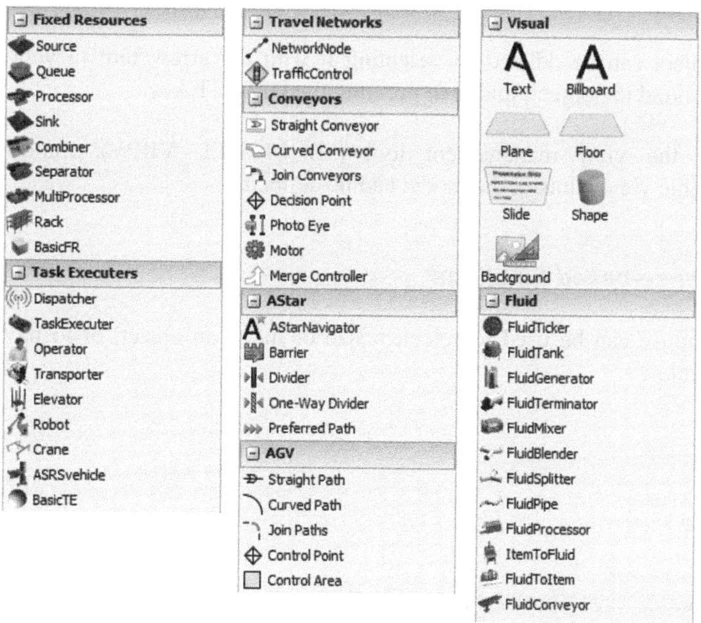

Figure 4.3. *The different libraries*

– AGV[3] (Automated Guided Vehicle):

- Straight Path: defines a straight path for an AGV;

- Curved Path: defines a curved path for an AGV;

- Join Paths: groups and joins AGV paths together.

– VISUAL: Library of tools created to expand the 3D model design and environment capabilities.

– ASTAR[3]: Library of tools based upon the A* algorithm.

– FLUID: Libraries dedicated to the generation, transport, storage, mixing and transformation of fluids.

The components forming the flows present in a model can be pallets, people, shapes, boxes, bottles, cardboard boxes, etc. All component types can be configured or imported into a model. Each of them entails a variety of information unique to them.

Objects can be inserted into the model by dragging them from the libraries or by clicking in model view, in this case right-clicking or pressing the ESC key returns you to standard mode (STANDARD MODE: white arrow in the upper horizontal toolbar).

An object can be deleted by selecting it with the arrow tool (a yellow frame appears around the object) and then pressing the DELETE key.

Using the view management tool (3D MODEL VIEW), multiple, fully customizable viewpoints in the model can be defined.

4.3.3. *Mouse-based functions*

The mouse can be used to select, resize or move an object, or to navigate the model window.

3 AGV, ASTAR and CONVEYOR are optional libraries which make new resources available. During the installation of FlexSim, the user can choose whether to include them or not.

Object Selection	
Single selection	Left-click
Multiple selection	CTRL key + left-click or SHIFT key + left-click
De-selection	SHIFT key + click on the background
Object Movement	
Translation in XY	Hold down left-click + move up/down or left/right
Translation in Z	Left-click + scroll up or down or Right-click + left-click and move up/down (*only works with certain configurations*)
Object Resizing	
Stretch/shrink along an axis	Left-click on the red cone (X axis), green cone (Y axis) or blue (Z axis)
Proportional stretch/ shrink (along all axes)	ALT key + left-click on one of the cones (red, green or blue)
Object Rotation	
Rotation about an axis	Right-click on a red cone (for X axis rotation), green cone (for Y axis rotation) or blue cone (for Z axis rotation)

Table 4.1. *The functionalities available with the mouse*

4.3.4. *Connections between objects*

The objects in a model are connected together using different ports: input, output and central ports.

The first two types of port connect fixed resources to each other and the central port sends orders, tasks or general messages to task executers. The central port is said to create a relative reference between two objects.

To create or destroy the different types of connection, we use tools from the main toolbar, or keyboard shortcuts (which are more practical):

– connection between objects (input/output): press **a** (and follow the flow direction) or click on the CONNECT OBJECTS tool (Figure 4.4);

– central connection: press **s** or click on the CONNECT CENTER PORTS tool (Figure 4.4);

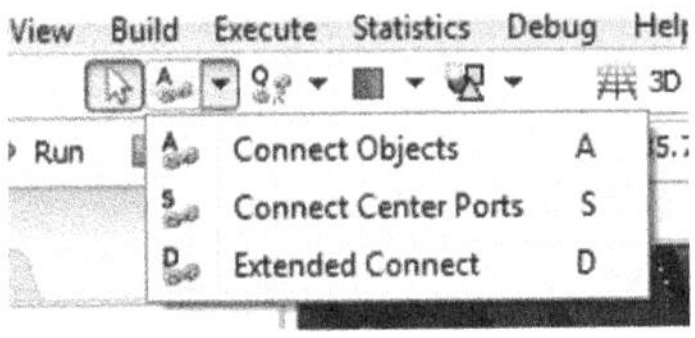

Figure 4.4. *The connection tools in the main toolbar*

– destroy connection between objects (input/output): press **q** (going in the same direction as for the creation) or click on the DISCONNECT OBJECTS tool (Figure 4.5);

– destroy central connection: press **w** or click on the DISCONNECT CENTER PORTS tool (Figure 4.5);

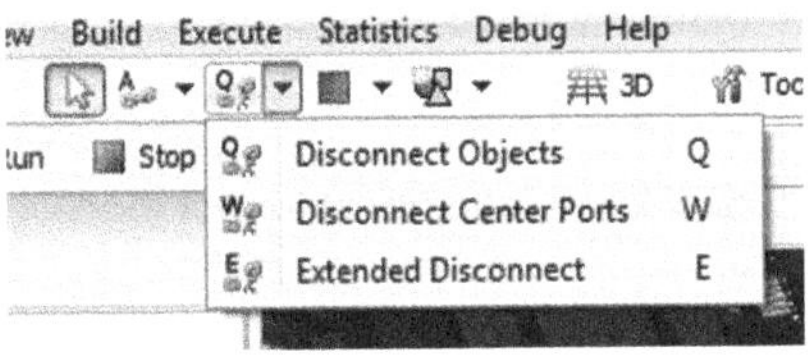

Figure 4.5. *Tools in the main toolbar for deleting links*

To create or delete a connection, select the starting object (a yellow box should appear), hold down the shortcut key (the cursor should turn into a chain), hold down the left mouse button and drag the cursor (this should draw out a yellow link) onto the final object (again a yellow box should appear) and release the click.

The connections between objects (inputs/outputs) are represented by small triangles situated at the end of the links, while central connections are represented by small squares.

4.4. Worked example

We will model a warehouse working with two systems. One part of the warehouse is automated with conveyors and three automated storage and retrieval system (ASRS) present, and the other is manual with pallets being transported by forklift truck operators.

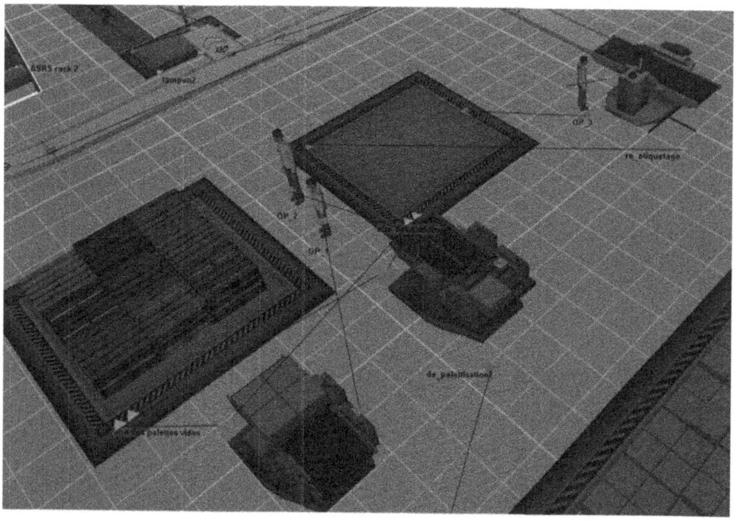

Figure 4.6. *Some examples of connections between objects and central connections (red and green triangles, red squares). For a color version of this figure, see www.iste.co.uk/reveillac/modeling3.zip*

4.4.1. *Description of the warehouse*

Figure 4.7 shows the plan of the warehouse.

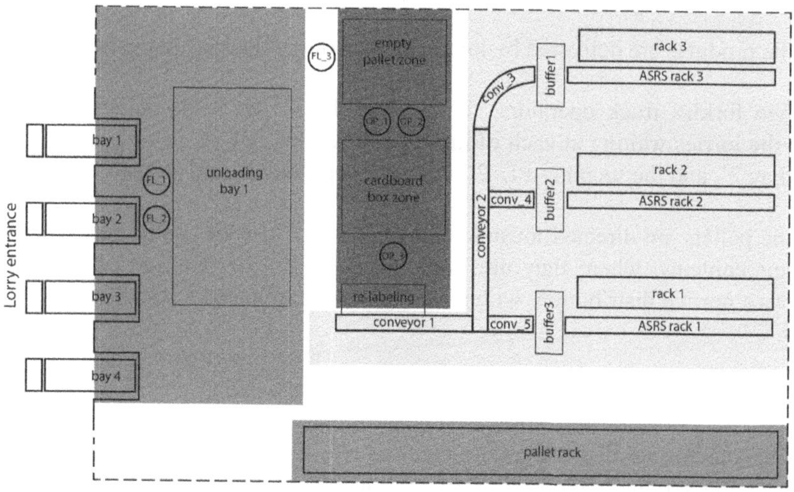

Figure 4.7. *The warehouse with its different zones and personnel. For a color version of this figure, see www.iste.co.uk/reveillac/modeling3.zip*

The warehouse comprises four zones:

– an unloading zone (green), containing four docking bays (bays 1–4) where lorries drop off their cargo, and an unloading bay;

– a zone for unloading pallets, storing empty pallets and storing cardboard boxes (purple) with an automated barcode-scanning and cardboard box relabeling machine;

– an automated storage zone with conveyors (conveyors 1–5), ASRS (ASRS racks 1–3) and product racks (racks 1–3) (yellow);

– a manual pallet storage zone (orange).

The warehouse personnel comprise three forklift truck operators (FL_1–3) and three operators (OP_1–3).

4.4.2. *Warehouse operation*

To make the operation of the warehouse more easily understandable, it will be broken down into three stages:

1) unloading;

2) unloading pallets and relabeling;

3) conveying and storing in racks.

4.4.2.1. *Unloading and storage*

The products are delivered by lorries at each of the docking bays (bays 1–4).

Two forklift truck operators, 'FL_1' and 'FL_2' fetch the pallets of products from the lorries waiting at each of the bays. The first, 'FL_1' is assigned to 'bay 1' and 'bay 2', and the second, 'FL_2', is assigned to 'bay 3' and 'bay 4'.

The pallets are directed toward 'unloading bay 1' or the 'pallet rack' according to their contents, where they are directly piled up. The duration of this activity follows a normal distribution, with a mean time of 6 min and a standard deviation of 100 s.

The unloading bay has a maximum capacity of 100 pallets.

The contents are distributed into eight pallet types (numbered 1–8). Only two of them are taken to the 'pallet rack' (large products).

The ‘pallet rack’ has 20 vertical rows over six levels. Each slot can store two pallets. Each new pallet is stored in the first available row and on the first available level.

The eight pallet types are apportioned as follows: 12, 18, 22, 15, 7, 3, 18 and 5% (making 100%). We assume that each pallet contains four cardboard boxes and that pallet types 4 and 8 (large cardboard boxes) are taken to the ‘pallet rack’.

4.4.2.2. *Unloading from pallets and relabeling*

The pallets placed in ‘unloading bay 1’ are taken one-by-one to the pallet unloading zone by a forklift truck operator, ‘FL_3’, where two operators, ‘OP_1’ and ‘OP_2’, remove the cardboard boxes and empty pallets. The duration of this operation follows a normal distribution with a mean time of 5 min and a standard deviation of 150 s.

The empty pallets are stored in the ‘empty pallet zone’ site, which can hold up to 200 pallets.

The cardboard boxes are stored in the ‘cardboard box zone’ site, which can hold up to 100 cardboard boxes.

One operator, ‘OP_3’ takes these cardboard boxes one-by-one to put them on a machine that automatically reads their barcodes and relabels them before passing them onto ‘conveyor 1’. The combination of these operations takes an average of 25 s per box.

4.4.2.3. *Transport and storing in racks*

‘Conveyor 1’, working with ‘conveyor 2’, delivers the products to one of three conveyors, ‘conv_3’, ‘conv_4’ or ‘conv_5’, according to the product type (of six different types).

The conveyors have a constant speed of 1 m per second.

Product types 1, 2 and 3 are directed toward ‘conv_3’, type 4 toward ‘conv_4’ and types 5 and 6 toward ‘conv_5’.

The proportions of each of the six product types are as follows:

– 15% are type 1;

– 12% are type 2;

– 10% are type 3;

– 37% are type 4;

– 15% are type 5;

– 11% are type 6.

Conveyors 3, 4 and 5 transport the cardboard boxes to three ASRS, through three buffer zones with a maximum capacity of 25, in this order:

– 'conv_3' *via* 'buffer1' toward 'ASRS rack3';

– 'conv_4' *via* 'buffer2' toward 'ASRS rack2';

– 'conv_5' *via* 'buffer3' toward 'ASRS rack1'.

The ASRS have a maximum speed of 2 m per second and take the boxes one-by-one.

Each of the three ASRS, 1–3, fills its corresponding rack, i.e. 'rack1', 'rack2' and 'rack3'.

The 'pallet rack' has eight vertical rows over nine levels. Each slot can take five cardboard boxes.

4.4.3. *Modeling stage 1*

Launch FlexSim then creates a new model by clicking on NEW MODEL. Keep the default units and start time suggested by the MODEL UNITS dialog box by clicking on OK.

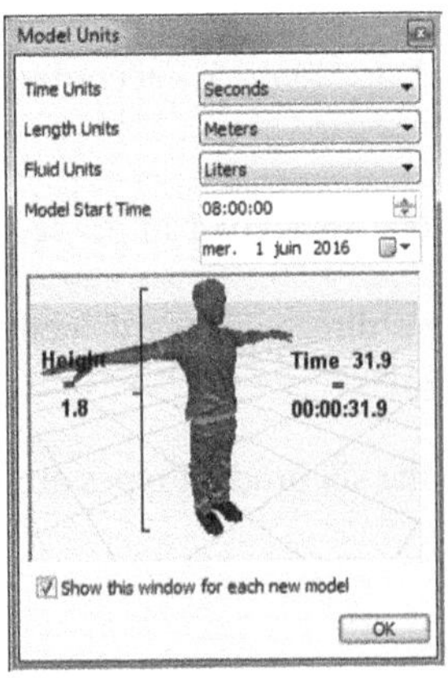

Figure 4.8. *The choice of units*

COMMENT 4.2.– By default, this dialog box appears each time a new model is created, until the SHOW THIS WINDOW FOR EACH NEW MODEL option has been unchecked.

4.4.3.1. *Creating the objects*

Place the objects in the model according to the plans above, as shown in Figure 4.9. This covers the unloading zone (green) and the pallet rack (orange).

The goods arrival bays 1–4 will be constructed from two SOURCE objects and one COMBINER object to generate more realistic pallet (PALLET) and cardboard box (BOX) subassemblies.

COMMENT 4.3.– These subassemblies are used to promote the use of functionalities more representative of the specific example. A SOURCE object directly generating cardboard boxes (BOX) would have sufficed.

All of the objects will be named in correspondence with the plans to provide consistency throughout the exercise. A double-click on each of the objects opens the properties window, whose top field contains the default name generated by FlexSim. This can be easily changed.

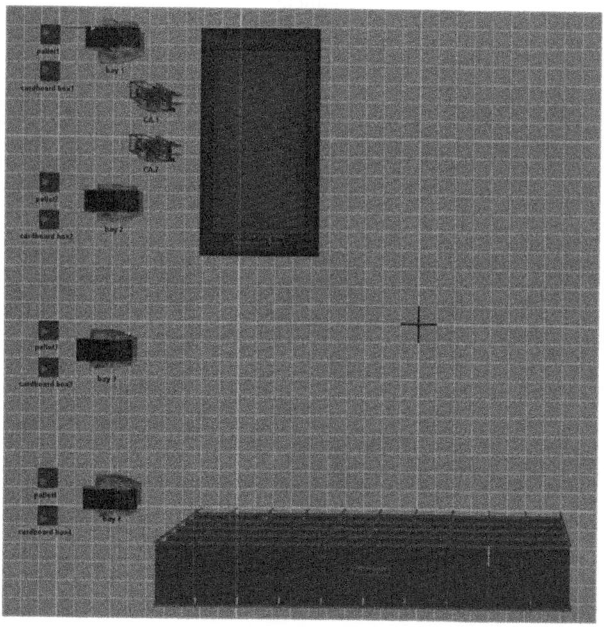

Figure 4.9. *The model with the objects for stage 1 of our example, as seen from above*

COMMENT 4.4.–

– In the following explanations, the names of the FIXED RESOURCES and the CONVEYORS are written in lower case and the names of the TASK EXECUTERS are written in upper case for clarity and to correspond to the plan in Figure 4.7.

– For object names, the dash and period characters are not accepted for certain properties, so an underscore will be used in their place.

– A right-click on the background of the model allows a choice from the MODEL VIEWS. A top-down view is probably easiest for positioning objects.

– The copy and paste functions can be very useful when multiple copies of an object exist.

Here is a list of the objects in the model (object type, number, name on the plans and any observations):

Object type	Number	Name on the plan (Figure 4.7)	Observations
SOURCE	8	pallet1 pallet2 pallet3 pallet4 box1 box2 box3 box4	These do not appear on the plan. A pallet source and a cardboard box source are placed in front of each COMBINER which represent bays 1–4.
COMBINER	4	bay 1 bay 2 bay 3 bay 4	To be placed side-by-side as on the plan (Figure 4.7)
QUEUE	1	unloading bay 1	Named 'unloading bay 1' on the object. Capacity: 100 pallets
RACK	1	pallet rack	20 slots over 6 levels
TRANSPORTER	3	FL_1 FL_2	2 forklift truck operators

Table 4.2. *The objects to be added to the example to complete stage 1*

4.4.3.2. *Connections*

Once the objects are present in the model, create the following connections:

– Connect the four sources, ‘pallet1’ to ‘pallet4’, to the four bays, ‘bay 1’ to ‘bay 4’, with connections of the type CONNECT OBJECTS (a).

– Connect the four sources, ‘box1’ to ‘box4’ to the four bays, ‘bay 1’ to ‘bay 4’ with connections of the type CONNECT OBJECTS (a).

– Connect the four bays, ‘bay 1’ to ‘bay 4’, to the ‘unloading bay’ zone with connections of the type CONNECT OBJECTS (a).

– Connect the four bays, ‘bay 1’ to ‘bay 4’, to the ‘pallet rack’ with connections of the type CONNECT OBJECTS (a).

– Connect forklift operator 1, ‘FL_1’, to bays ‘bay 1’ and ‘bay 2’ with connections of the type CONNECT CENTER PORTS (s).

– Connect forklift operator 2, ‘FL_2’, to bays ‘bay 3’ and ‘bay 4’ with connections of the type CONNECT CENTER PORTS (s).

Figure 4.10. *Our model after adding all of the connections. For a color version of this figure, see www.iste.co.uk/reveillac/modeling3.zip*

4.4.3.3. *Configuration*

Now, we have to configure all of the objects.

At this stage, if you launch the simulation by clicking on the RUN icon in the auxiliary toolbar, after setting a speed on the RUN SPEED slider, you should see the unloading bay fill up. Objects 'bay 1' to 'bay 4' should transport boxes (default item). This confirms the correct implementation of your model.

Click on the STOP symbol, then on the RESET symbol to return the model to its initial setup.

Do not forget to save the model (FILE menu, SAVE...). The default extension for FlexSim models is '*fsm*'.

Configure the sources, 'pallet1' to 'pallet4', in their properties windows (double-click on the object):

– SOURCE tab: FLOWITEM CLASS (type of item generated), PALLET;

– TRIGGERS tab: in ON EXIT, DATA, add properties (green + symbol), SET ITEM TYPE BY PERCENTAGE, SET ITEMTYPE AND COLOR (Figure 4.11).

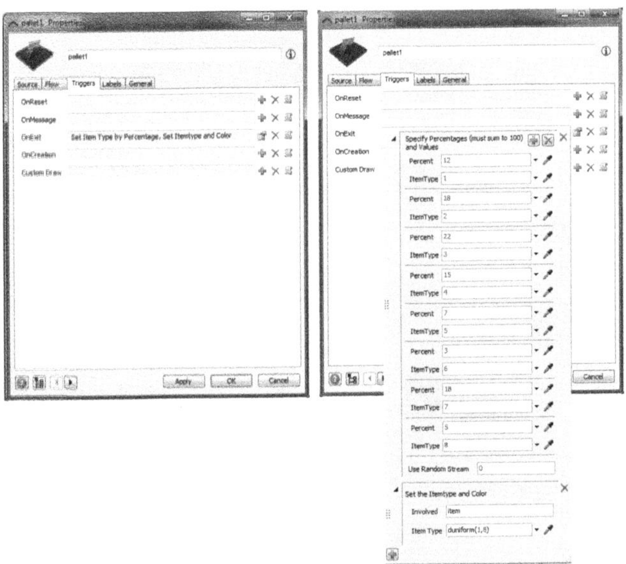

Figure 4.11. *The TRIGGERS tab in the configuration window for a pallet source (on the left). The two properties and their attributes (on the right), 12, 18, 22, 15, 7, 3, 18, and 5% as well as eight items of different colors, 'duniform(1,8)'*

COMMENT 4.5.–

– Under the TRIGGERS tab, the symbol in the form of a green + adds a property and the red cross symbol deletes one.

– The symbol in the form of a hand pointing at text shows the properties.

– The small white triangle in front of a property opens up the property contents to show its attributes.

– For a COMBINER object, output port 1 is always the container item, in this case, the pallet.

– Make sure to confirm all choices by clicking on APPLY and/or OK to close the window.

– The eight colors are chosen by FlexSim – the correspondence of object types to percentages and colors has been arbitrarily chosen here.

– The total sum of the percentages for the SET ITEM TYPE BY PERCENTAGE property must equal 100 (MUST SUM TO 100).

Configuration for the sources, 'box1' to 'box4', in their properties windows (double-click on the object):

– SOURCE tab: FLOWITEM CLASS (type of item generated), BOX;

– SOURCE tab: ARRIVAL AT TIME 0 option checked.

Configuration – for the combiners, 'bay 1'–'bay 4':

– PROCESSTIMES tab: PROCESS TIME, STATISTICAL DISTRIBUTION, NORMAL, MEAN = 360, STDDEV = 100 (all quantities are expressed in seconds, here giving 6 min = 360 s). The field must contain: '`normal(360, 100, 0)`';

– COMBINER tab: COMBINER MODE, PACK (cardboard boxes side-by-side);

– COMBINER tab: TARGET QUANTITY = 4 (four boxes per pallet);

– FLOW tab: SEND TO PORT, PORT BY CASE, CASE 4 and 8 toward port 2;

– FLOW tab: USE TRANSPORT option checked (using forklift truck operators), see Figure 4.12.

Configuration for the 'unloading bay 1' zone:

– QUEUE tab: MAXIMUM CONTENT, 100;

– QUEUE tab: ITEM PLACEMENT, STACK INSIDE QUEUE.

Configuration of the 'pallet rack':

– RACK tab: MAXIMUM CONTENT, 240 (20 × 6 × 2);

– RACK tab: PLACE IN BAY, FIRST AVAILABLE BAY and MAXIMUM ITEM PER CELL, 2;

– RACK tab: PLACE IN LEVEL : FIRST AVAILABLE LEVEL and MAXIMUM ITEM PER CELL, 2;

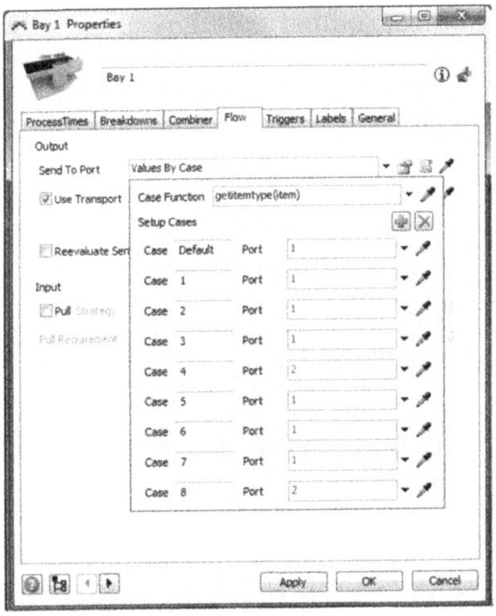

Figure 4.12. *The parameters under the FLOW tab for the bays*

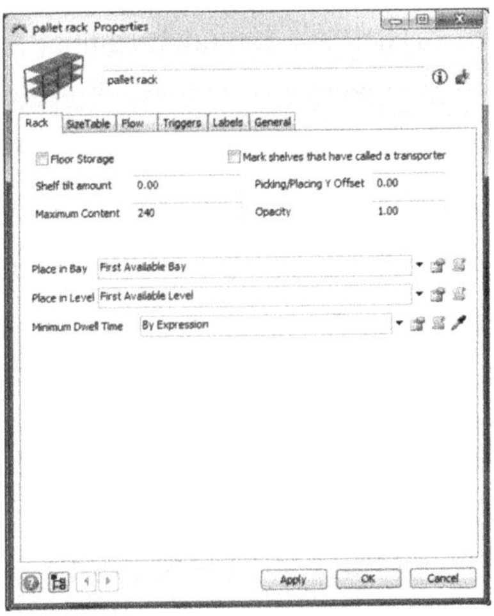

Figure 4.13. *The RACK tab in the configuration window for the 'pallet rack' object*

– SIZE TABLE tab: NUMBER OF BAYS, 20;

– SIZE TABLE tab: NUMBER OF LEVELS, 6;

– SIZE TABLE tab: WIDTH OF BAYS, 2 m;

– SIZE TABLE tab: HEIGHT OF LEVELS, 1 m.

COMMENT 4.6.– Clicking on the APPLY BASIC SETTINGS button under the SIZE TABLE tab in a rack properties dialog box applies the parameters input for NUMBER OF BAYS, NUMBER OF LEVELS, WIDTH OF BAYS and HEIGHT OF LEVELS. To create custom racks (number of levels depends on columns), the parameters under the ADVANCED zone must be configured.

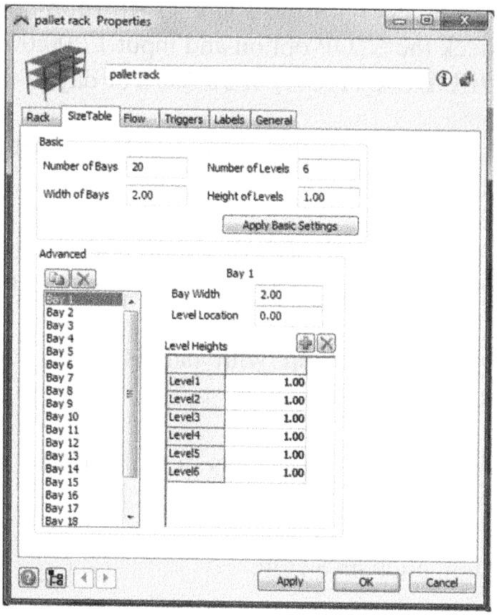

Figure 4.14. *The SIZE TABLE tab of the configuration window for the 'pallet rack' object*

Configuration of the 'empty pallet zone':

– QUEUE tab: MAXIMUM CONTENT, 100;

– QUEUE tab: ITEM PLACEMENT, STACK INSIDE QUEUE.

Configuration of the 'cardboard box zone':

– QUEUE tab: MAXIMUM CONTENT, 400;

– QUEUE tab: ITEM PLACEMENT, STACK INSIDE QUEUE.

Do not forget to save the model (FILE menu, SAVE…)

4.4.3.4. *Testing the model*

Before testing the model, we will configure the simulation duration.

Go into the FILE menu, select MODEL SETTINGS and define:

– the start time (MODEL START TIME): 8:00 am;

– the date: 1st June 2016;

– the TIME FORMAT: HH:mm:ss, for two-digit hour in 24-h format.

In the auxiliary toolbar, click on the RUN TIME drop-down menu, check BY DATE AND TIME, check the STOP option and input 17:00:00. You should have an overall duration (MODEL DURATION) of 32 400 s (1 day of 9 h).

Click on RUN in the auxiliary toolbar to launch the simulation. The speed can be varied using the RUN SPEED slider.

Pallets of eight different colors and boxes representing cardboard boxes should circulate around the model, and the two forklift operators should start moving.

The unloading bay should fill up with pallets of six different colors, each carrying four cardboard boxes.

The pallet rack should fill up with pallets of two different colors, each carrying four cardboard boxes.

To stop the simulation, click on STOP.

The RESET button allows the simulation to be relaunched from the start, while taking account of any changes that might have been made.

COMMENT 4.7.– Every FlexSim window contains a green, circular icon with a question mark, which can be used to obtain help from the user manual.

4.4.4. *Modeling stage 2*

We will now model and configure the depalletizing and relabeling zone in the model (purple).

4.4.4.1. *Creating the objects*

Add the following elements to the model:

Object type	Number	Name on the plan (Figure 4.7)	Observations
QUEUE	2	empty pallet zone cardboard box zone	Capacity: 200 pallets Capacity: 100 boxes
SEPARATOR	2	'de_palletizing1' and 'de_palletizing2' Although these operations do not appear on the plan, they must be included in the de-palletizing zone (purple) to formalize the operators' work.	
PROCESSOR	1	re-labeling	Named 're_ labeling' on the object.
TRANSPORTER	1	FL_3	1 forklift truck operator
OPERATOR	3	OP_1 OP_2 OP_3	3 operators

Table 4.3. *The objects to be added for stage 2 of the example*

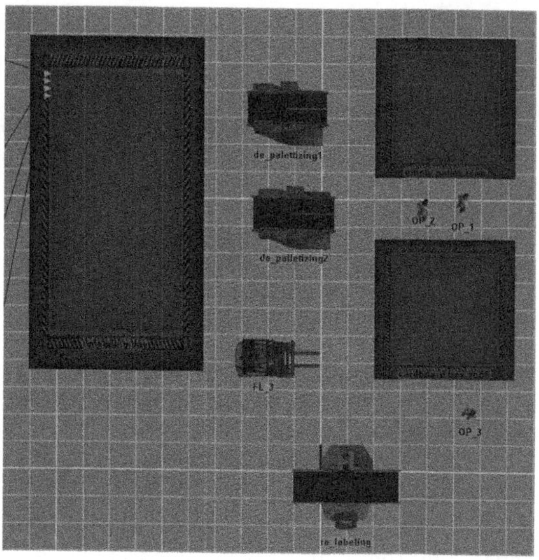

Figure 4.15. *The right-hand part of the model with the objects that have just been added (three operators, one forklift operator, two separators, one processor and two queues)*

4.4.4.2. *Connections*

With these new objects in the model, create the following connections:

– Connect the 'unloading bay 1' zone to the two depalletizing separators, 'de_palletizing1' and 'de_palletizing2', with connections of type CONNECT OBJECTS (a).

– Connect the third forklift truck operator, 'FL_3', to the 'unloading bay 1' zone with a connection of type CONNECT CENTER PORTS (s).

– Connect the first operator, 'OP_1', to the 'de-palletizing1' separator with a connection of type CONNECT CENTER PORTS (s).

– Connect the second operator, 'OP_2', to the 'de_palletizing2' separator with a connection of type CONNECT CENTER PORTS (s).

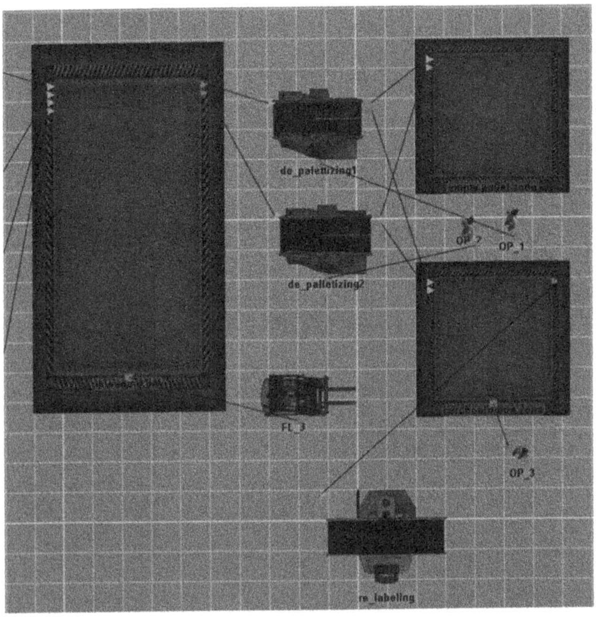

Figure 4.16. *The new connections in the model for stage 2. For a color version of this figure, see www.iste.co.uk/reveillac/modeling3.zip*

– Connect each of the depalletizing separators, 'de_palletizing1' and 'de_palletizing2', first to the 'empty pallet zone' site (on port 1) then to the 'cardboard box zone' site (on port 2) with connections of type CONNECT OBJECTS (a).

– Connect the 'cardboard box zone' site to the barcode scanning and relabeling site, 're_labeling', with a connection of type CONNECT OBJECTS (a).

– Connect the 'OP_3' operator to the 'cardboard box zone' site with a connection of type CONNECT CENTER PORTS (s).

4.4.4.3. *Configuration*

Configuration of the 'unloading bay 1' site:

– Under the FLOW tab: check the USE TRANSPORT option.

Configuration of the 'de_palletizing1' and 'de_palletizing2' separators:

– PROCESSTIMES tab: PROCESS TIME, STATISTICAL DISTRIBUTION, NORMAL, MEAN = 240, STDDEV = 150 (the quantities are all expressed in seconds, here with 5 min = 300 s). The field must contain: '`normal(300, 150, 0)`';

– FLOW tab: check the USE TRANSPORT option (use operators);

– FLOW tab: SEND TO PORT, DEFAULT SEPARATOR OPTION, PORT FOR CONTAINER, 1 and PORT FOR CONTENTS, 2. This configuration just needs to be confirmed, as it should already be in place by default. Port 1 (CONTAINER) should lead to the empty pallet zone, and port 2 (CONTENTS) to the cardboard box zone.

Configuration of the 'empty pallet zone' site:

– QUEUE tab: MAXIMUM CONTENT, 200;

– QUEUE tab: ITEM PLACEMENT, STACK INSIDE QUEUE.

Configuration of the 'cardboard box zone' site:

– QUEUE tab: MAXIMUM CONTENT, 100;

– QUEUE tab: ITEM PLACEMENT, STACK INSIDE QUEUE;

– FLOW tab: check the USE TRANSPORT option.

Configuration of the barcode scanning and relabeling machine, 're_labeling':

– PROCESSOR tab: PROCESS TIME, STATISTICAL DISTRIBUTION, EXPONENTIAL, SCALE = 25;

– TRIGGERS tab: in ON EXIT, DATA, add the SET ITEMTYPE BY PERCENTAGE property and SET ITEMTYPE AND COLOR (Figure 4.17).

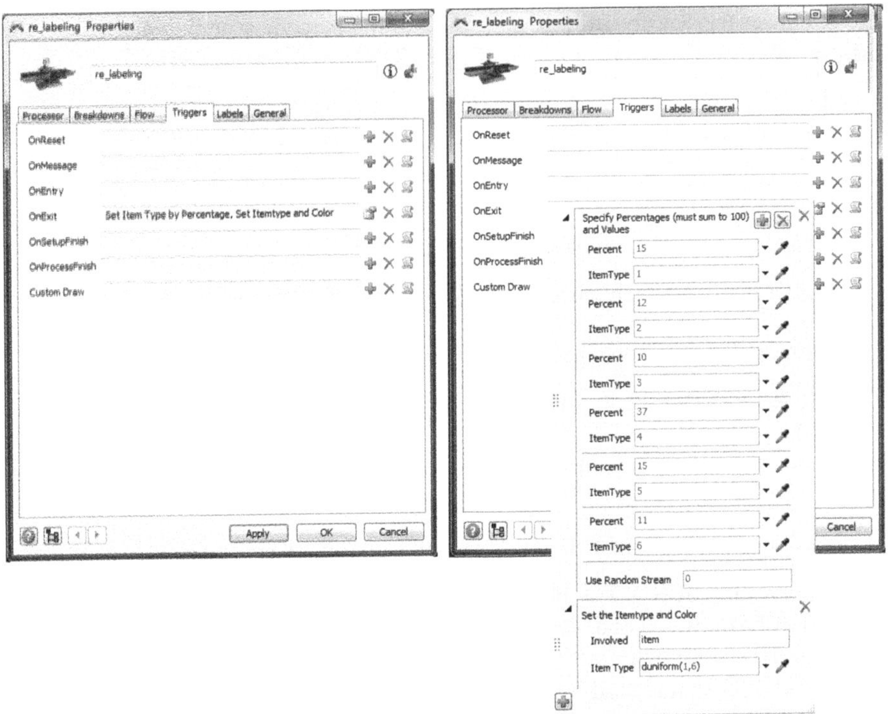

Figure 4.17. *The SET ITEMTYPE BY PERCENTAGE (6 items: 15, 12, 10, 37, 15 and 11%) and SET ITEMTYPE AND COLOR ('duniform(1,6)') properties of the 're_labeling' processor*

COMMENT 4.8.– Every FlexSim window features a green circular icon with a question mark, which can be used to obtain help from the user manual.

Do not forget to save the model (FILE menu, SAVE…)

4.4.4.4. *Testing the model*

Launch the simulation by clicking on RUN in the auxiliary toolbar. The speed can be varied using the RUN SPEED slider.

In parallel with the operations carried out by stage 1, you should see the forklift truck operator, 'FL_3', and the two operators, 'OP_1' and 'OP_2' start working.

The empty pallet zone should fill up with pallets of six different colors, and the cardboard box zone should fill up with boxes arriving in batches of four.

COMMENT 4.9.– During the simulation, if the pallets or cardboard boxes overflow the racks, adjust their size.

To check that the barcode scanning and relabeling machine correctly differentiates between the various box types, you can complete the following:

– Stop the simulation by clicking on STOP.

– Add a QUEUE object after the 're-labeling' processor.

– Connect the 're-labeling' processor to this QUEUE object with a connection of type CONNECT OBJECTS (a).

– Rerun the simulation after clicking on the RESET button.

You should see cardboard boxes of six different colors pile up on the QUEUE object that has just been added. This confirms the correct functioning of the model.

Delete this object once the test is complete.

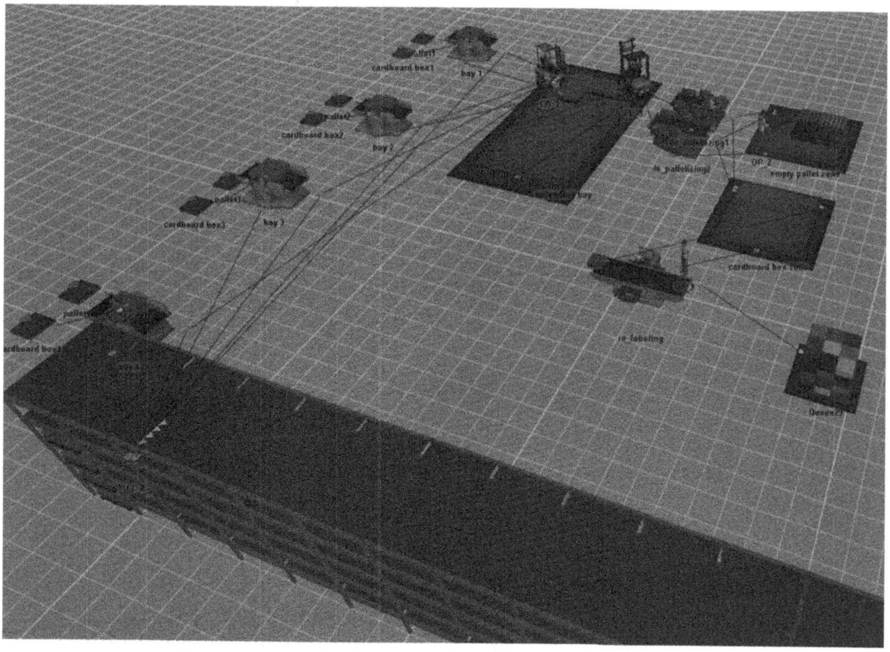

Figure 4.18. *The model after stage 2, during simulation. The QUEUE object added to check correct functioning of the 're-labeling' processor can be seen on the right. For a color version of this figure, see www.iste.co.uk/reveillac/modeling3.zip*

4.4.5. *Modeling stage 3*

The cardboard boxes of the products will now be conveyed to the three automated storage robots (ASRS) according to their types. They will then be stored in the three racks.

We will now model the yellow zone on the plan (Figure 4.7).

4.4.5.1. *Creating the objects*

Add the following elements to the model:

Object type	Number	Name on the plan (Figure 4.7)	Observations
STRAIGHT CONVEYOR	4	conveyor 1 conveyor 2 conveyor 3 conveyor 4	To be placed side-by-side, as on the plan (Figure 4.7) Check direction
CURVED CONVEYOR	1	conv_3	Check direction
QUEUE	3	buffer1 buffer2 buffer3	3 buffer zones
RACK	3	rack 1 rack 2 rack 3	8 slots over 9 levels
ASRS	3	ASRS rack 1 ASRS rack 2 ASRS rack 3	3 automated storage robots
DECISION POINT	4	– DP1, to be placed on conveyor 1; – DP2, to be placed on conveyor 3; – DP3, to be placed on conveyor 4; – DP4, to be placed on conveyor 5. Although these points do not appear on the plan, they are required to manage the distribution of the boxes to the conveyors according to their types.	

Table 4.4. *The stage 3 objects*

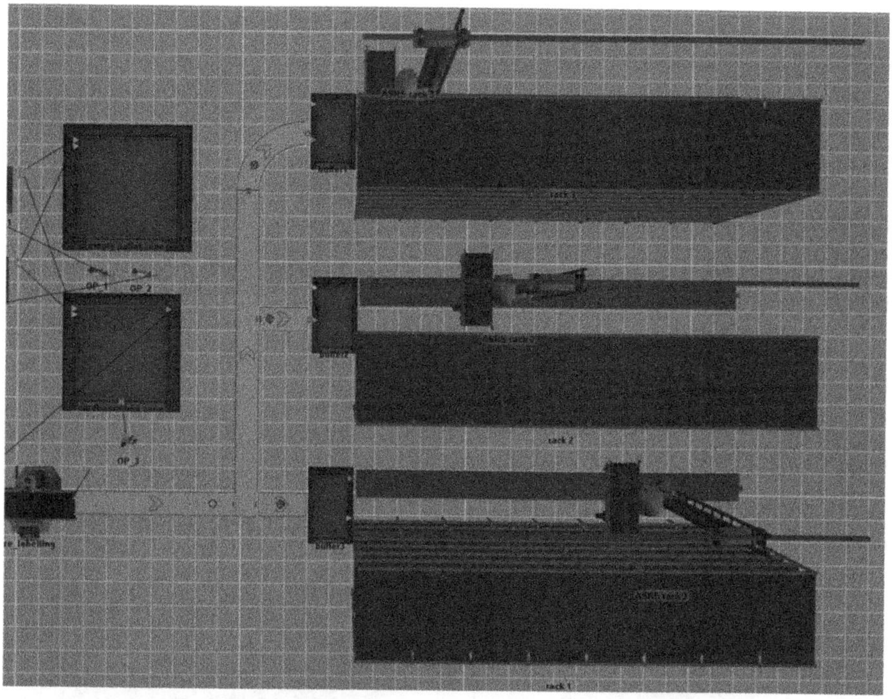

Figure 4.19. *The objects added to the model in stage 3 (five conveyors, four decision points, three queues, three ASRS and three racks). For a color version of this figure, see www.iste.co.uk/reveillac/modeling3.zip*

4.4.5.2. *Connections*

Create the following connections from the new model objects:

– Connect the 're_labeling' barcode scanning and relabeling machine to 'conveyor 1' with a connection of type CONNECT OBJECTS (a).

– Connect the decision point on 'conveyor 1' to the decision point on 'conv_3'.

– Connect the decision point on 'conveyor 1' to the decision point on 'conv_4'.

– Connect the decision point on 'conveyor 1' to the decision point on 'conv_5'.

– Connect the conveyor, 'conv_3' to the buffer zone, 'buffer1', with a connection of type CONNECT OBJECTS (a).

– Connect the conveyor, 'conv_4' to the buffer zone, 'buffer2', with a connection of type CONNECT OBJECTS (a).

– Connect the conveyor, ‘conv_5’ to the buffer zone, ‘buffer3’, with a connection of type CONNECT OBJECTS (a).

– Connect the buffer zone, ‘buffer1’, to the storage rack, ‘rack 3’, with a connection of type CONNECT OBJECTS (a).

– Connect the buffer zone, ‘buffer2’, to the storage rack, ‘rack 2’ with a connection of type CONNECT OBJECTS (a).

– Connect the buffer zone, ‘buffer3’, to the storage rack, ‘rack 1’, with a connection of type CONNECT OBJECTS (a).

– Connect the storage robot, ‘ASRS rack 3’, to the buffer zone, ‘buffer1’, with a connection of type CONNECT CENTER PORTS (s).

– Connect the storage robot, ‘ASRS rack 2’, to the buffer zone, ‘buffer2’, with a connection of type CONNECT CENTER PORTS (s).

– Connect the storage robot, ‘ASRS rack 1’ to the buffer zone, ‘buffer3’, with a connection of type CONNECT CENTER PORTS (s).

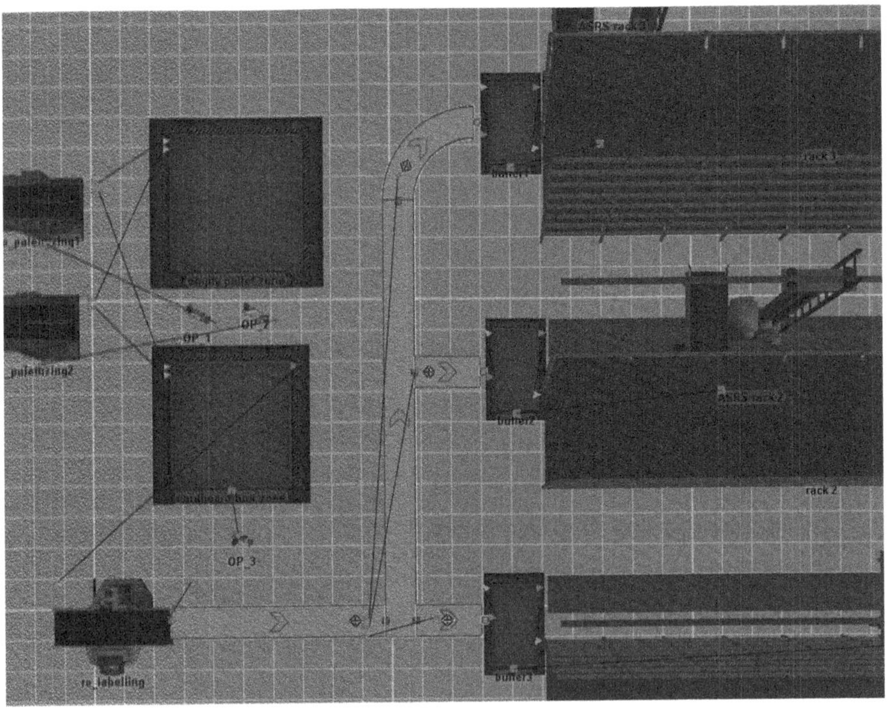

Figure 4.20. *The stage 3 object connections. For a color version of this figure, see www.iste.co.uk/reveillac/modeling3.zip*

Configuration of the decision point, 'DP1', on 'conveyor 1':

– DECISION POINT TYPE tab: ON ARRIVAL, SET ITEM BY CASE and add the six cases (1, 2 and 3 toward 'conv_3', i.e. port 1; 4 toward 'conv_4', i.e. port 2; and 5, 6 toward 'conv_5', i.e. port 3).

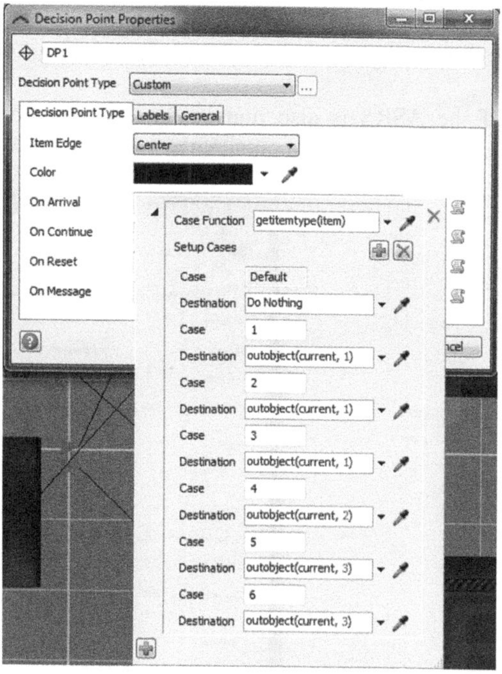

Figure 4.21. *The properties of the decision point assigned to 'conveyor 1'*

Configuration of buffer zones, 'buffer1', 'buffer2' and 'buffer3':

– QUEUE tab: MAXIMUM CONTENT 25;

– FLOW tab: check the USE TRANSPORT option.

Configuration of the racks, 'rack 1', 'rack 2' and 'rack 3':

– RACK tab: MAXIMUM CONTENT, 360 ($8 \times 9 \times 5$);

– RACK tab: PLACE IN BAY, FIRST AVAILABLE BAY and MAXIMUM ITEM PER CELL, 8;

– RACK tab: PLACE IN LEVEL: FIRST AVAILABLE LEVEL and MAXIMUM ITEM PER CELL, 8;

– SIZE TABLE tab: NUMBER OF BAYS, 8;

– SIZE TABLE tab: NUMBER OF LEVELS, 9;

– SIZE TABLE tab: WIDTH OF BAYS, 2 m;

– SIZE TABLE tab: HEIGHT OF LEVELS, 1 m;

Configuration of the conveyors is not required, their default SPEED is 1 m/s.

Configuration of the ASRS is also unnecessary because they already have a MAX SPEED of 2 m/s and always take one item at a time (CAPACITY).

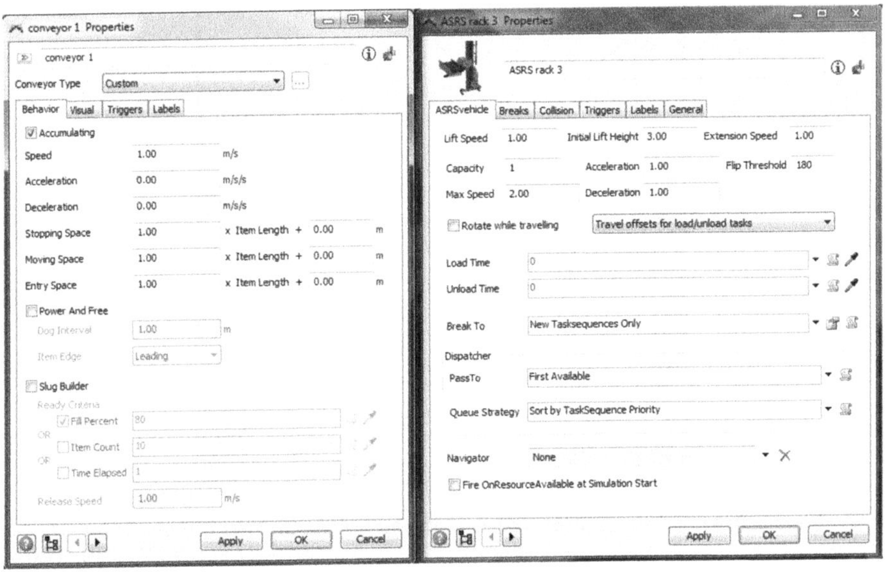

Figure 4.22. *The default parameters for a conveyor (left) and an ASRS (right)*

4.4.5.3. *Dashboards and analysis*

At this point, we will add two dashboards to give us a quantitative view of various elements of the simulation.

Click on the DASHBOARDS icon in the main toolbar and select ADD A DASHBOARD.

A new window with a DASHBOARD tab should open on the right-hand side of the model. Clicking on this, a new list of tools should appear in the LIBRARY window on the left-hand side of the model.

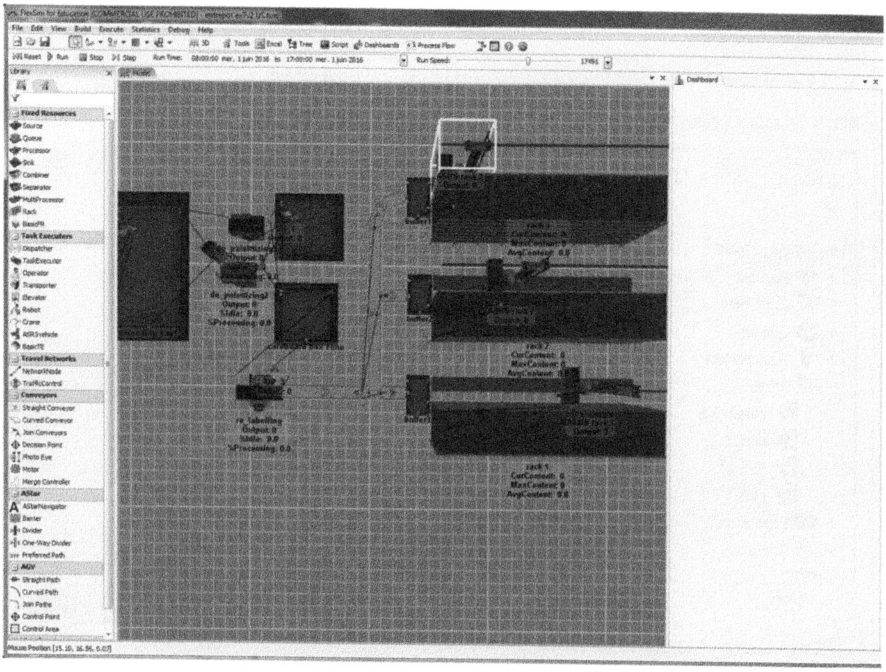

Figure 4.23. *FlexSim with our model (in the center) and the LIBRARY and DASHBOARD windows (on the left and right, respectively)*

Drag the AVERAGE CONTENT object from the LIBRARY window on the left toward the DASHBOARD window.

The object may need to be rescaled to fit correctly.

A STAT CHART PROPERTIES dialog box associated with this object should open.

Under the OBJECTS tab, click on the green + symbol, choose SELECT OBJECTS, then the + symbol next to RACKS and click on SELECT.

Change the name of the DASHBOARD, instead of AVERAGE CONTENT (top field in the dialog box) input 'Mean rack contents'.

Then, click on OK.

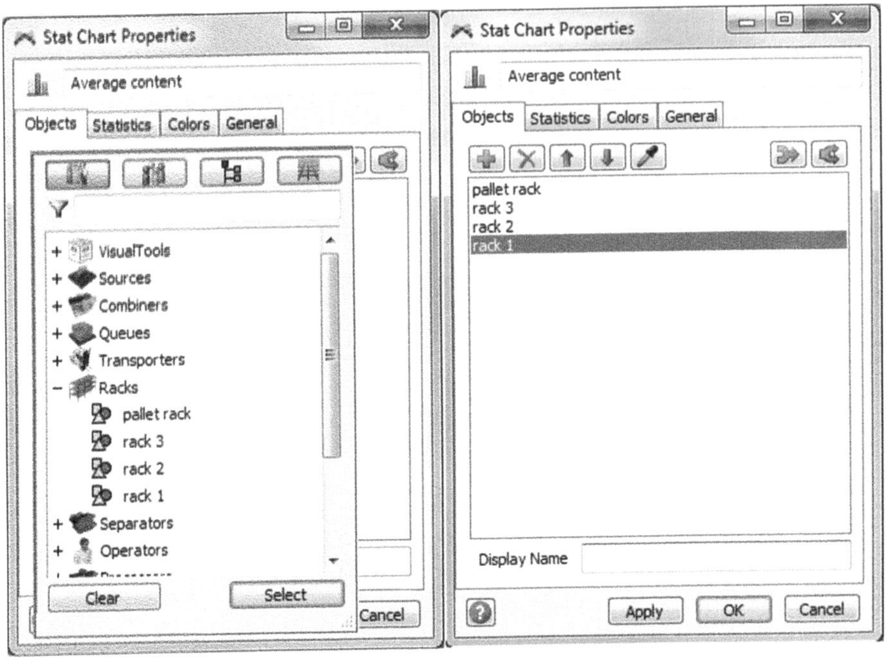

Figure 4.24. *The STAT CHART PROPERTIES dialog box before and after selecting the four racks*

You should obtain a dashboard like the one shown in Figure 4.25.

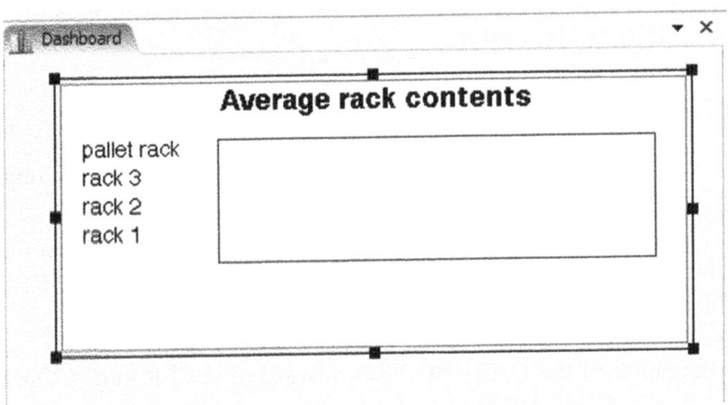

Figure 4.25. *The dashboard that will indicate, using a horizontal bar chart, the mean contents (in number of pieces) of each rack*

Add a second dashboard for visualizing the activity of the fixed resources, 'bay 1 to 4' (COMBINER), 'de-palletizing_1 and 2' (SEPARATOR) and 're-labeling' (PROCESSOR), which incorporate the worktimes of the personnel, operators and forklift truck operators (see the note in section 4.4.4.6).

Drag a STATE PIE dashboard into the DASHBOARD window.

Add to it the objects, bays '1'–'4', 'de-palletizing '1' and '2' and 're-labeling', and give it the title, 'Resource activity' (Figure 4.26).

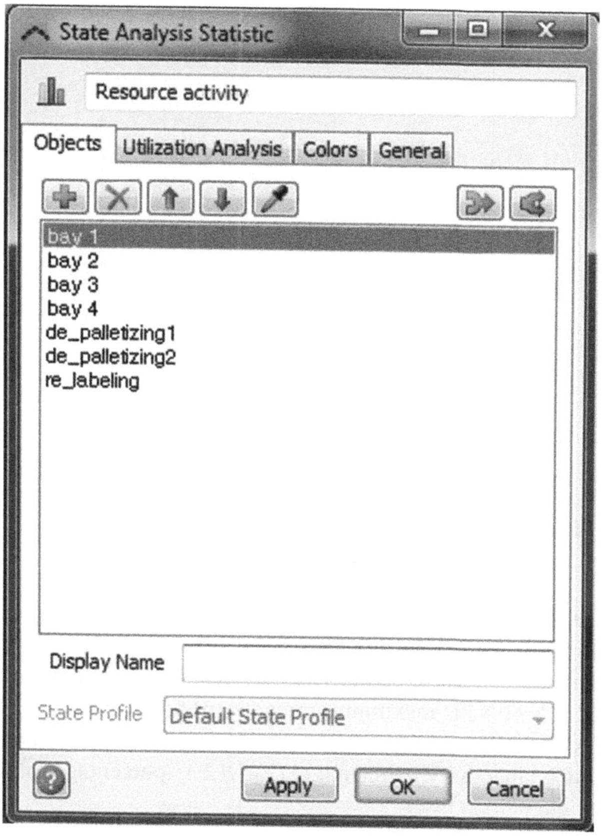

Figure 4.26. *The objects associated with the second dashboard, named 'Resource activity'*

This second dashboard will display seven pie charts, so make sure that it is large enough to display these easily.

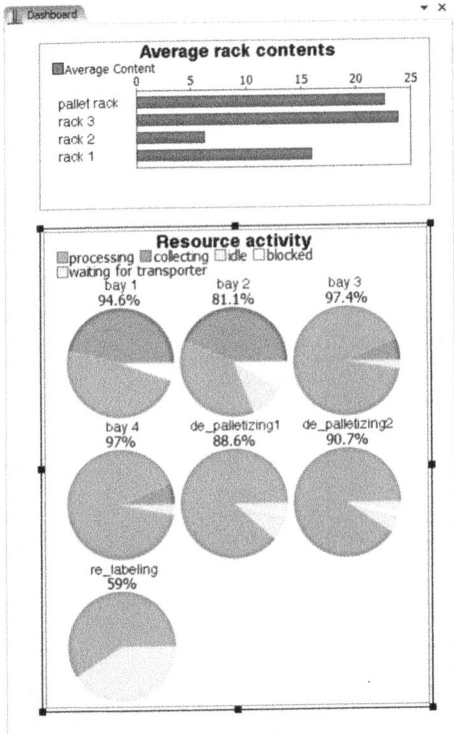

Figure 4.27. *The two dashboards during simulation, displaying a bar chart and seven pie charts. For a color version of this figure, see www.iste.co.uk/reveillac/modeling3.zip*

COMMENT 4.10.– By scrolling down the SHOW NAMES list in the VIEW SETTINGS window and selecting SHOW NAMES AND STATS, different statistical data related to each object can be seen during the simulation: current contents (CURCONTENT), maximum contents (MAXCONTENT), mean waiting time (AVGSTAYTIME), number of items output (OUTPUT), activity percentage (%PROCESSING), inactivity percentage (%IDLE), percentage of time blocked (%BLOCKED), etc.

4.4.5.4. *Testing the model*

The model for this first example is now finished.

Run the simulation. The dashboards should display the desired graphics, all varying throughout the simulation. The ASRS should activate and store the cardboard boxes according to type, symbolized by color in our model.

Rack 1 should contain three cardboard boxes of three different colors, rack 2 should contain boxes of just one color and rack 3 should contain boxes of two different colors.

By changing the properties and rerunning the simulation after pressing RESET, the simulation, and hence graphics, should change too.

For example, change the process durations for the bays (COMBINER) and the relabeling machine, and the conveyor speed.

You can try:

– PROCESS TIME: '`normal(200, 80, 0)`' for the bays;

– PROCESS TIME: '`exponential(0.0, 15, 0)`' for the re-labeling machine;

– SPEED: '2m/s' for all conveyors.

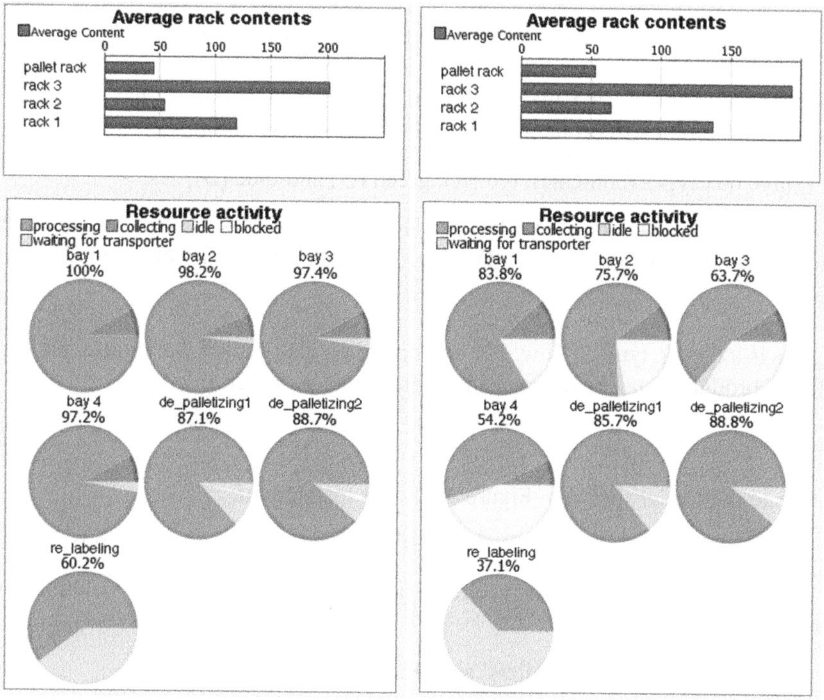

Figure 4.28. *The two dashboards for: the initial parameters (left) and the modified parameters (right). For a color version of this figure, see www.iste.co.uk/reveillac/modeling3.zip*

COMMENT 4.11.– The travel, loading and unloading times have not been taken into account for each of the task executers in this model (OPERATOR, TRANSPORTER and ASRS) in order to simplify the simulation. They are assumed to be included in the process durations (TIME PROCESS) of the fixed resources (COMBINER, SEPARATOR and PROCESSOR).

4.5. Detailed flow and task executer management

The basic application of some of the FlexSim resources has been demonstrated in the previous example. Some functionalities will now be explored in detail in this section.

The aim of this section is to provide some solutions to problems that are commonly encountered in warehouses. They can serve as inspiration for developing your own models.

4.5.1. *Generation of containers with several types of content*

The aim is to use a COMBINER to randomly generate pallets (containers) carrying different boxes (contents) according to certain conditions:

– three box types (contents): red (R), green (G) and blue (B);

– one pallet can contain: 1R, 1R-1V, 1R-1V-1B, 2R-1V-1B, 2R-2V-1B, 2R-2V-2B, 3R-2V-2B, 3R-3V-2B, 3R-3V-3B, 3R, making 10 possible combinations overall.

This is a classic type of flow. For example, in a distribution warehouse, pallets of different products must be dispatched according to customer orders.

4.5.1.1. *Model and connections*

Create the model shown in Figure 4.29. It is made up of:

– Four SOURCEs: 'Pallet', 'Red_boxes', 'Green_boxes' and 'Blue_boxes';

– Three RACKs: 'Red rack', 'Green rack' and 'Blue rack' (five columns over eight levels);

– Two QUEUEs: 'Stock pallets' and 'Full pallets';

– One COMBINER: 'Crea_pallet'.

All the connections are of type CONNECT OBJECTS (a).

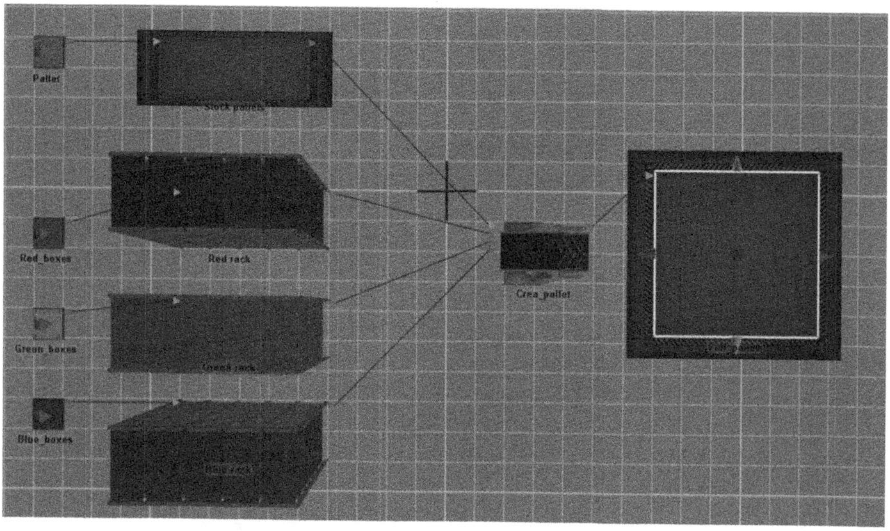

Figure 4.29. *The model and all of its connected resources. For a color version of this figure, see www.iste.co.uk/reveillac/modeling3.zip*

4.5.1.2. *Configuring the objects and creating a table*

To make the model's appearance match the specifications described above, the color of each object can be changed in the COLOR field under the GENERAL tab of the PROPERTIES dialog box (double-click on the object).

The racks will contain boxes of a single color, so the sources must generate them correctly.

Open the 'Red_boxes' source properties and under the SOURCE tab, scroll down the FLOWITEM CLASS list and select GO TO FLOWITEM BIN…

In the window that opens (the FLOWITEM BIN tab), click on the green + icon and select NEW BASIC FLOWITEM.

Set the dimensions of the brown box that appears in the GENERAL PROPERTIES window to 0.4, 0.4 and 0.3 (for example). This item will become the box, so name it and change its color accordingly (Figure 4.31).

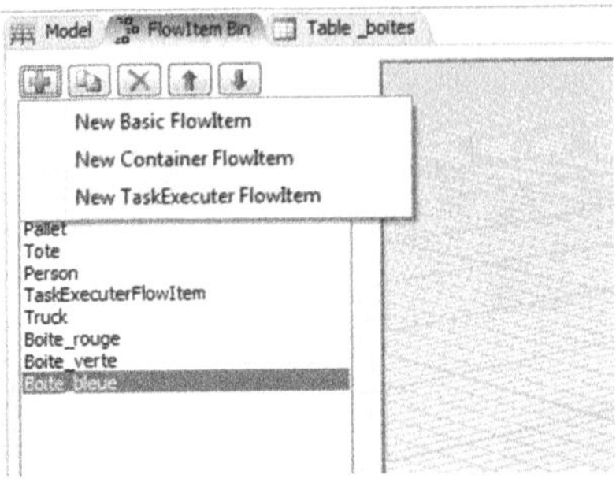

Figure 4.30. *Add a new item*

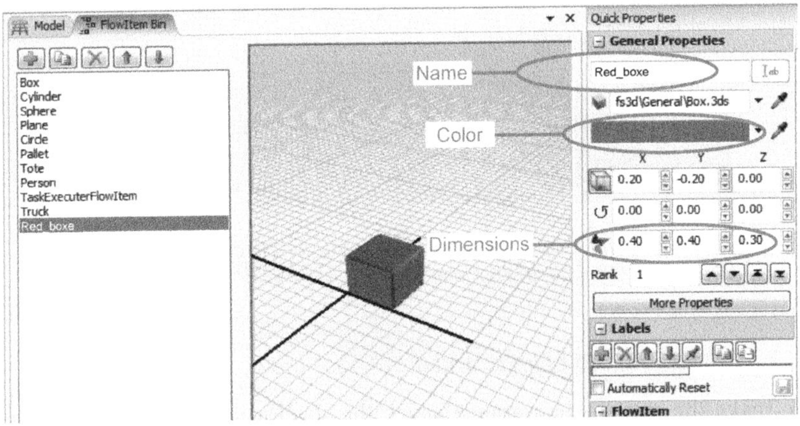

Figure 4.31. *Configuring the new item (here 'Red_boxes'). For a color version of this figure, see www.iste.co.uk/reveillac/modeling3.zip*

Repeat the operation for the two other sources, green and blue.

COMMENT 4.12.– You can duplicate an existing item using the icon resembling two sheets of paper, between the green + symbol and the red cross under the FLOWITEM BIN tab.

Click on the MODEL tab or close the FLOWITEM BIN.

For the 'Pallet' source, under the TRIGGERS tab of the properties window, add a function to the field ONEXIT: SET ITEMTYPE. Define 10 item types: '`duniform(1, 10)`'. These will be the 10 combinations mentioned above, in section 4.5.1.

We will now turn to the 'Crea_pallet' combiner.

First, we will create a table containing the different possible combinations for the three box types.

Go into the VIEW menu and select TOOLBOX. In the TOOLBOX window, click on the green + icon and select GLOBAL TABLE to add a new table.

Figure 4.32. *The TOOLBOX window and its add tool*

Create a table with three rows and 10 columns, name it 'Box_table' and fill it out as shown in Figure 4.33.

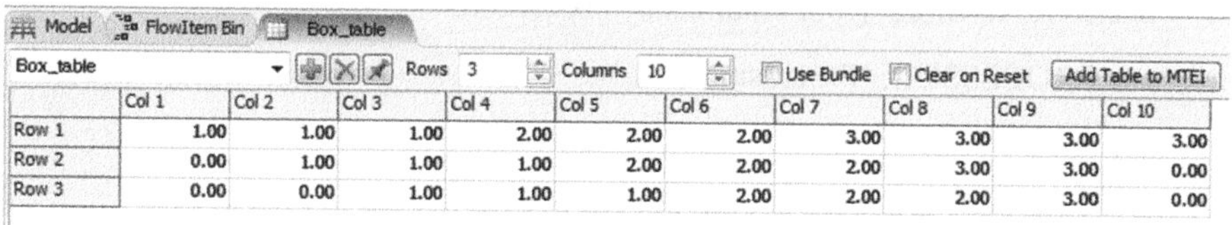

	Col 1	Col 2	Col 3	Col 4	Col 5	Col 6	Col 7	Col 8	Col 9	Col 10
Row 1	1.00	1.00	1.00	2.00	2.00	2.00	3.00	3.00	3.00	3.00
Row 2	0.00	1.00	1.00	1.00	2.00	2.00	2.00	3.00	3.00	0.00
Row 3	0.00	0.00	1.00	1.00	1.00	2.00	2.00	2.00	3.00	0.00

Figure 4.33. *The table of the 10 possible box combinations (section 4.5.1)*

Return to the model by clicking on the MODEL tab or closing the BOX_TABLE window.

Open the 'Crea_pallet' combiner properties, go to the TRIGGERS tab and select UPDATE COMBINER COMPONENT LIST for the ONENTRY field.

Close the properties window by clicking on OK to confirm the modifications.

4.5.1.3. *Testing the model*

Click on RUN to launch the simulation.

Pallets carrying combinations of red, green and blue boxes should be randomly distributed along the 'Full_pallets' queue.

This sequence is random because the choice of item number is defined by the 'Pallet' source from its ONEXIT trigger.

4.5.2. *A fixed resource for task executers*

In this example, we will fill a storage rack with items of three different colors. Two operators will then collect the items to transport them to a processing unit, which will take them to a dispatch zone.

One of the specifications thus entails sharing one fixed resource, the rack, between two operators.

Furthermore, we instruct one of the operators to transport items of a single color, and the other to transport the two remaining item colors.

4.5.2.1. *Model and connections*

Create the model shown in Figure 4.34, made up of:

– one SOURCE: 'Supply';

– one RACK: 'Store' (four columns by six levels);

– one PROCESSOR: 'Processing';

– one QUEUE: 'Dispatch';

– two OPERATORs: 'OP_1' and 'OP_2'.

All of the fixed resources are connected by connections of type CONNECT OBJECTS (a).

The 'Store' rack is connected to the two operators, 'OP_1' (in central port 1) and 'OP_2' (in central port 2), with connections of type CONNECT CENTER PORTS (s).

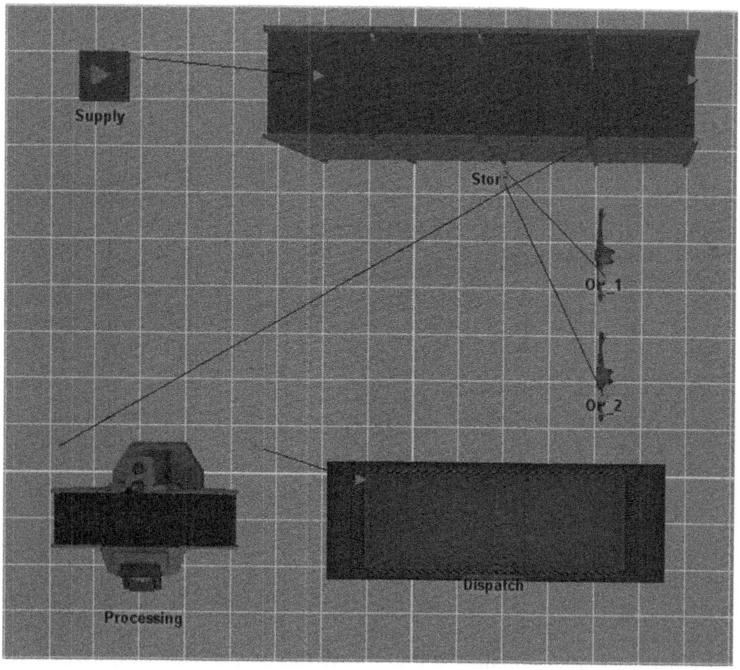

Figure 4.34. *The model and its resources and connections. For a color version of this figure, see www.iste.co.uk/reveillac/modeling3.zip*

4.5.2.2. *Configuring the objects*

For the 'Supply' source, go into TRIGGERS, then place SET ITEMTYPE AND COLOR in the ONEXIT field. Three types should already be specified by default: '`duniform(1,3)`'.

Change the rack dimensions in the SIZE TABLE tab: NUMBER OF BAYS, 4; NUMBER OF LEVELS, 6.

Go to the FLOW tab, check USE TRANSPORT and in the corresponding field select MULTIPLE TRANSPORT OPTIONS, specifying port 1 for item 1 and port 2 for items 2 and 3. These settings will control the tasks allocated to the operators, 'OP_1' for one item and 'OP_2' for two items.

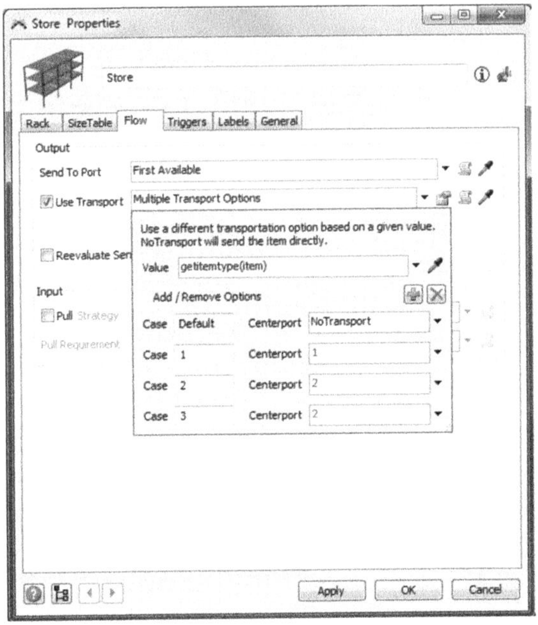

Figure 4.35. *Configuring the FLOW tab of the rack properties*

For the other objects, the 'Processing' processor and the 'Dispatch' queue, keep the default properties.

4.5.2.3. *Testing the model*

Launch the simulation by clicking on RUN.

The operator, 'OP_1', should supply the 'processing' unit with boxes of a single color and the operator, 'OP_2', should supply two different colors of boxes.

4.5.3. *Shared task executers*

We consider two stores, the first containing three different product types, and the second containing just one type.

These products should be transported toward two processing units by three operators and a forklift truck operator.

The operators are able to transport products of type 1 and 2, and the forklift truck operator product types 3 and 4.

Products 1 and 2 are taken to the first processing unit, and the others to the second processing unit.

Products 1 and 2 are stocked in a specific place. The same is true for products 3 and 4.

4.5.3.1. *Model and connections*

Create the model shown in Figure 4.36, with:

– one SOURCE: 'Supply_1';

– two PROCESSORs: 'Processing_1' and 'Processing_2';

– four QUEUEs: 'Store_1', 'Store_2', 'Product_1' and 'Product_2';

– three OPERATORs: 'OP_1', 'OP_2' and 'OP_3';

– one TRANSPORTER: 'FL_1';

– one DISPATCHER: 'Dispatcher'.

All of the fixed resources are connected with connections of type CONNECT OBJECTS (a). The queues, 'Store_1' and 'Store_2', are both connected to the processors, 'Processing_2', in port 2 for the first queue and port 1 for the second.

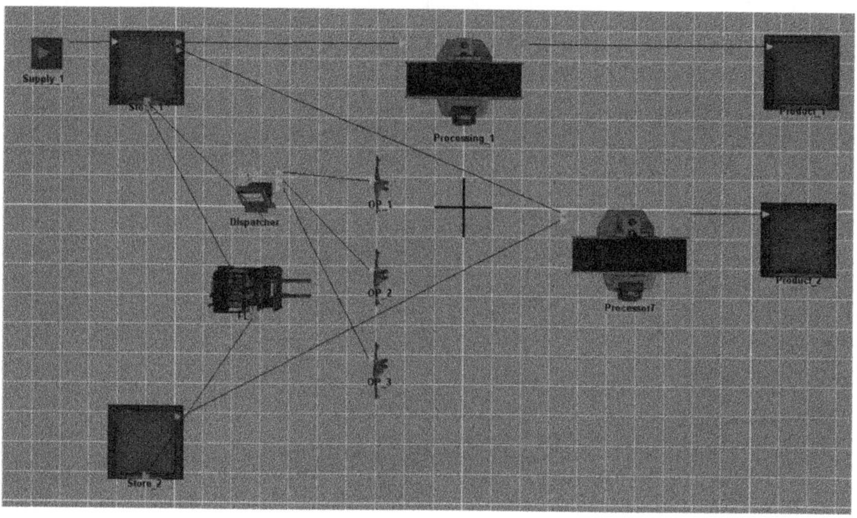

Figure 4.36. *The model and its connections. The four resources and the dispatcher are visible. For a color version of this figure, see www.iste.co.uk/reveillac/modeling3.zip*

The 'Dispatcher' is connected to the queue, 'Store_1' (in central port 1), and the forklift truck operator, 'FL_1', is connected to the 'Store_1' queue (in central port 1) and to the 'Store_2' queue (in central port 1) with connections of type CONNECT CENTER PORTS (s).

The 'Dispatcher' is also connected to the three operators, 'OP_1', 'OP_2' and 'OP_3', with three connections of type CONNECT OBJECTS (a).

4.5.3.2. *Configuring the objects*

For the 'Supply_1' source, under the TRIGGERS tab in the ONEXIT field, select SET ITEMTYPE AND COLOR and keep the default option that generates three items, '`duniform(1,3)`'.

For the 'Store_1' queue, under the FLOW tab in the SEND TO PORT field, select PORT BY CASE. Check USE TRANSPORT and select the MULTIPLE TRANSPORT OPTIONS option. All of the parameters are shown in Figure 4.37.

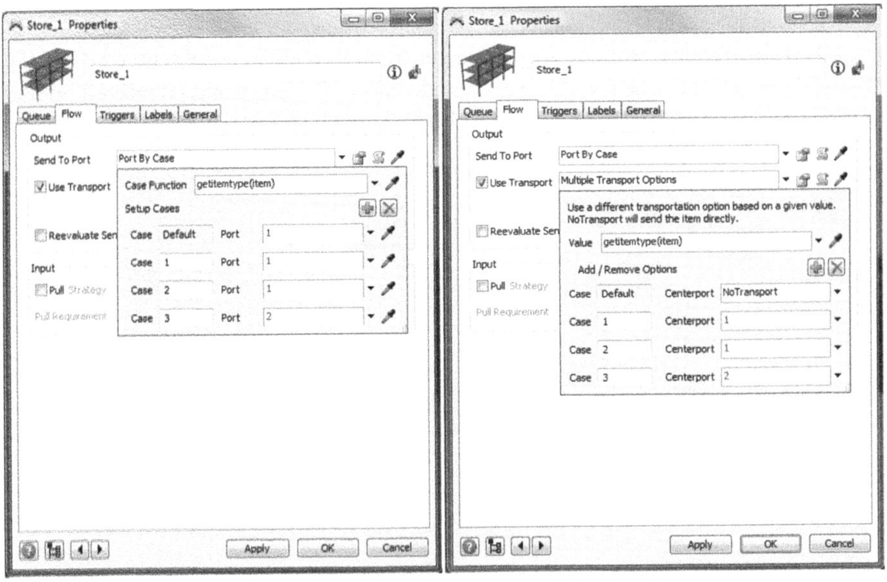

Figure 4.37. *The parameters in the FLOW tab of the 'Store_1' queue*

The PORT BY CASE option directs the items toward the two output ports, items 1 and 2 toward port 1 and item 3 toward port 2.

The MULTIPLE TRANSPORT OPTION directs items toward the dispatcher or the forklift truck operator. Items 1 and 2 go to the 'Dispatcher' to be dealt with by the operators, while item 3 goes to the forklift truck operator, 'FL_1'.

For the 'Dispatcher', select ROUND ROBIN for the PASS TO field under the DISPATCHER tab. This option will use the three operators in turn.

For the 'Store_2' queue, we will select a strategy which will fill the queue with 50 boxes immediately after initialization.

Go to the ONRESET field, select SEND MESSAGE and `current` for the FROM field (default option) and `current` for the TO field (needs inputting). These parameters transmit a message to the object itself, here the 'Store_2' queue.

We set the queue to create 50 type 1 items, in this case BOXes, upon receiving this message. Select CREATE FLOWITEMS for the field ONMESSAGE with 50 in QUANTITY TO CREATE and 1 in FLOWITEM BIN RANK (Figure 4.38).

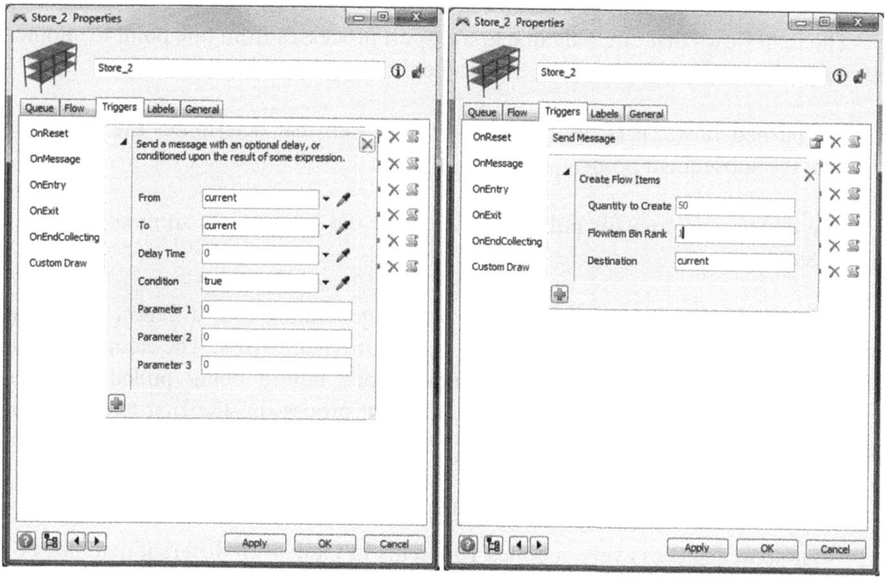

Figure 4.38. *The parameters under the TRIGGERS tab of the 'Store_2' queue*

The processors, 'Processing_1' and 'Processing_2', do not require configuration. Their default properties can be kept.

The same is true for the queues, 'Product_1' and 'Product_2'.

4.5.3.3. *Testing the model*

Click on RUN to launch the simulation.

The 'Store_2' queue should instantly fill with 50 boxes of the same color, and the 'Store_1' queue should constantly fill itself with blocks of three different colors.

The forklift truck operator, 'FL_1', should carry blocks of two different colors to the 'Processing_2' processor, from 'Store_1' and 'Store_2'.

The three operators 'OP_1', 'OP_2' and 'OP_3' should, in turn, collect boxes of two different colors and take them to the 'Processing_1' processor.

The two queues 'Product_1' and 'Product_2' should each fill up with blocks of two different colors.

4.5.4. *Pulled and pushed flows and more*

The item flows that circulate due to transport processes from one point to another are subject to two main strategies, they can either be pushed or pulled.

For pushed flows, it is the initial processing step that determines the arrival of items at the subsequent point.

The opposite is true for pulled flows, where the later points summon items as they need them.

In the following example, we have a source generating four different items that pushes them onto two processors for an operation lasting 10 s. The items are then placed in a communal intermediate storage zone before being pulled by three processors for an operation lasting 60 s, the first processing the first two types of item and the two others processing types 3 and 4, respectively.

To add a bit more detail to this example, the following constraints are added:

– our source successively supplies 100 items in total, with different quantities of each type;

– initially, item type 1: quantity = 40;

– followed by item type 2: quantity = 30;

– followed by item type 3: quantity = 10;

– and finally item type 4: quantity = 20.

4.5.4.1. *Model and connections*

Create the model shown in Figure 4.39, with the following objects:

– one SOURCE: 'Products';

– five PROCESSORs: 'Operation_1' to 'Operation_5';

– one QUEUE: 'Storage zone';

– one SINK: 'Exit'.

All of the fixed resources are connected with connections of type CONNECT OBJECTS (a).

Make sure to create two connections from the 'Storage zone' to 'OP_3'.

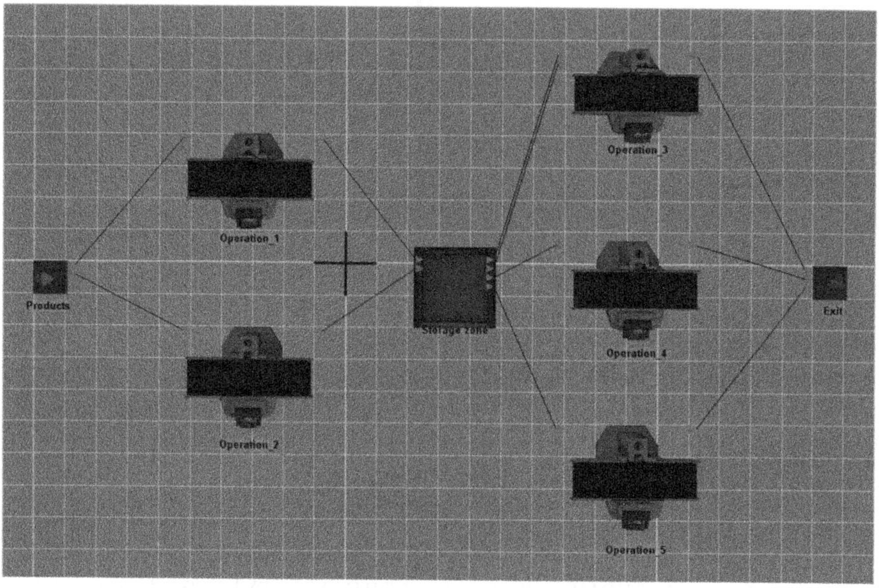

Figure 4.39. *The model for managing pushed and pulled flows. For a color version of this figure, see www.iste.co.uk/reveillac/modeling3.zip*

4.5.4.2. *Configuring the objects*

For the 'Products' source, go to the FLOW tab, select PORT BY CASE in the SEND TO PORT field, and configure the four cases to obtain: CASE 1 – PORT 1; CASE 2 – PORT 1; CASE 3 – PORT 2; CASE 4 – PORT 2. These options will direct items of types 1 and 2 toward output port 1 and items 2 and 3 toward output port 2.

Under the TRIGGERS tab of the 'Products' source, in the ONCREATION field, select SET COLOR BY CASE.

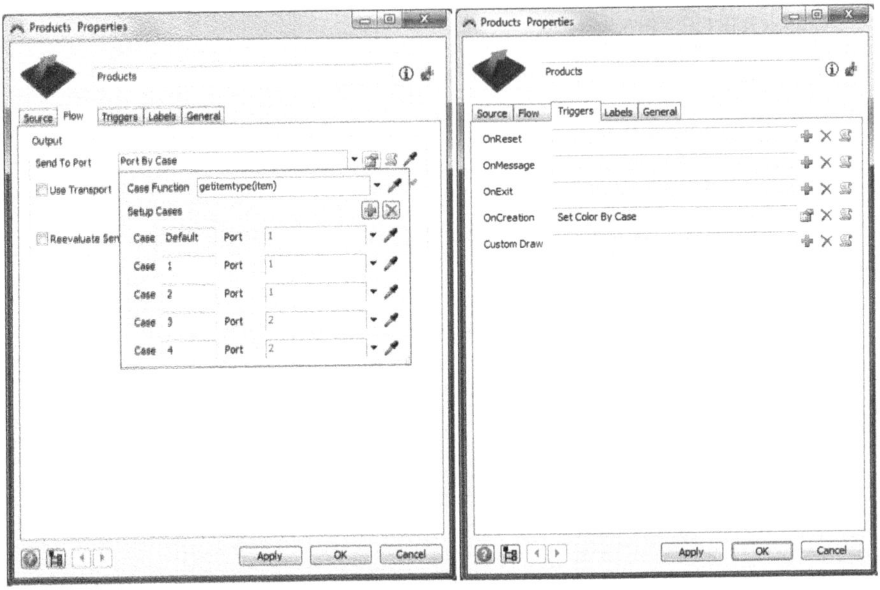

Figure 4.40. *The configuration of the FLOW and TRIGGERS tabs for the 'Products' source*

COMMENT 4.13.– The choice of the ONCREATION field is obligatory in this case, indeed if we had chosen ONEXIT, the item types would not have been taken into account in the redirection by the SEND TO PORT of the FLOW tab.

To take account of the quantities to be generated, we will specify them under the SOURCE tab with an ARRIVALS TABLE.

Select ARRIVAL SEQUENCE for the ARRIVAL STYLE field, input 4 into NUMBER OF ARRIVALS then click on REFRESH ARRIVALS to update the table beneath.

Fill in each row of the table with the ITEMNAME, the ITEM TYPE and the QUANTITY:

– ARRIVAL1: P1; 1; 40;

– ARRIVAL2: P2; 2; 30;

– ARRIVAL3: P3; 3; 10;

– ARRIVAL4: P4; 4; 20.

This table (Figure 4.41) also shows that we have named the items ‘P1’–‘P4’, which could be useful for later processes.

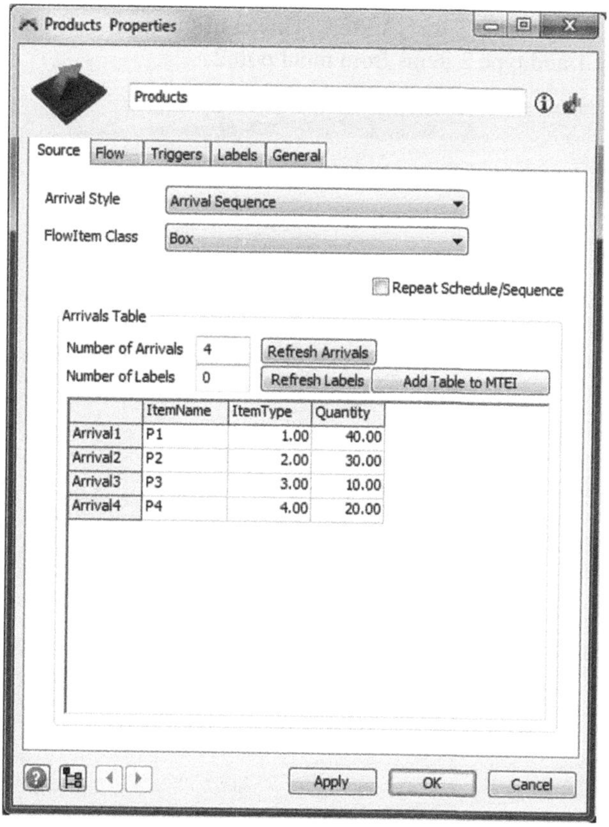

Figure 4.41. *The parameters under the SOURCE tab of ‘Products’ with its arrivals table*

Go to the PROCESSOR tab of each of the processors, ‘Operation_3’, ‘Operation_4’ and ‘Operation_5’, and input the duration of the process in the PROCESS TIME field, here 60 (seconds).

The duration does not have to be input for the processors, ‘Operation_1’ and ‘Operation_2’, because they already have a default duration of 10 s.

The flows in a processor are pushed by default, so to obtain pulled flows, this must be specified in the FLOW tab.

Go to the FLOW tab of the processor, 'Operation_3', check PULL under INPUT and select ITEMTYPE BY PORT in the PULL REQUIREMENT field, with PORT 1 for TYPE 1 and PORT 2 for TYPE 2. This configuration will pull type 1 items from input port 1 and type 2 items from input port 2.

Figure 4.42. *Configuration for pulling two items from two input ports of the 'Operation_3' processor*

Go to the FLOW tab of each of the processors, 'Operation_4' and 'Operation_5', check PULL under INPUT and select SPECIFIC ITEMTYPE for the PULL REQUIREMENT field, with a value of 3 for 'Operation_4' and a value of 4 for 'Operation_5'. Only type 3 items will be processed by 'Operation_4', and only type 4 items by 'Operation_5'.

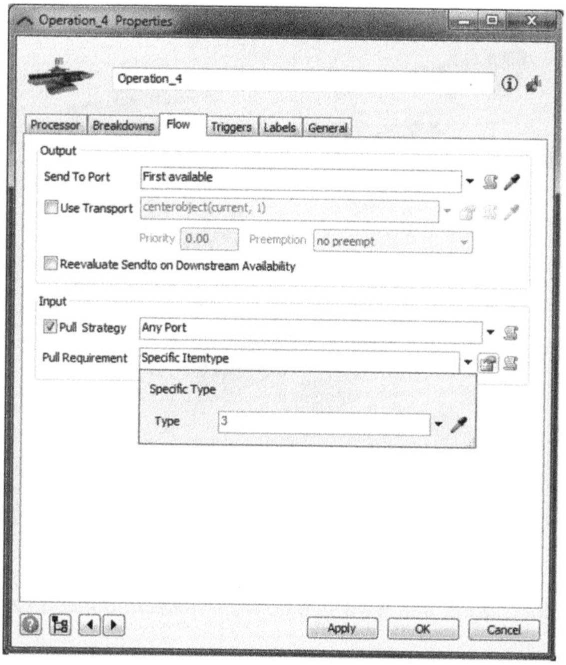

Figure 4.43. *Configuration for pulling item 3 on input port 1 of the 'Operation_4' processor*

4.5.4.3. *Testing the model*

After launching the simulation, the storage zone should successively fill up with 40 type 1 items, then 30 type 2 items, followed by 10 type 3 items and finish with 20 type 4 items.

The 'Operation_1' processor should process items of types 1 and 2.

The 'Operation_2' processor should process items of types 2 and 3.

The 'Operation_3' processor should process items of types 1 and 2, which it pulls from the 'Storage zone' queue.

The 'Operation_4' processor should process type 3 items, which it pulls from the 'Storage zone' queue.

The 'Operation_5' processor should process type 4 items, which it pulls from the 'Storage zone' queue.

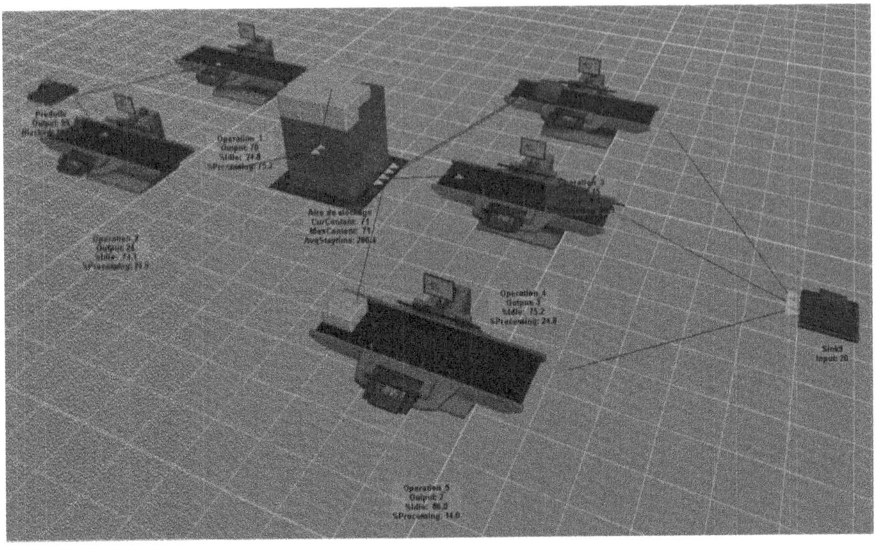

Figure 4.44. *Our model during simulation. For a color version of this figure, see www.iste.co.uk/reveillac/modeling3.zip*

The processors ‘Operation_3’, ‘Operation_4’ and ‘Operation_5’ work in parallel, drawing items from the ‘Storage zone’ queue, which is used as a buffer zone since the process durations of the upstream (average 10 s) and downstream (average 60 s) processors are different.

4.5.5. *Naming items*

It is often useful to be able to name the items carried by a flow through a process, to make it more realistic.

To illustrate this principle, we will model a very simple example.

A processing unit is supplied with items from two sources, with items named ‘Product_1’ for the first source and ‘Product_2’ from the second.

‘Product_1’ items represent 60% of the items produced, with ‘Product_2’ items representing the remaining 40%.

The processing unit can process five items sequentially.

After processing, these items are placed in two different storage sites by two operators.

Each of the operators handles only one type of product.

4.5.5.1. *Model and connections*

Create the model shown in Figure 4.45, with:

– two SOURCEs: 'P1 source' and 'P2 source';

– one PROCESSOR: 'Processing';

– two OPERATORs: 'OP_1' and 'OP_2';

– two QUEUEs: 'P1 store' and 'P2 store'.

All of the fixed resources are connected to each other with connections of type CONNECT OBJECTS (a).

The task executers, 'OP_1' and 'OP_2', are connected to the 'Processing' processor with connections of type CONNECT CENTER PORTS (s).

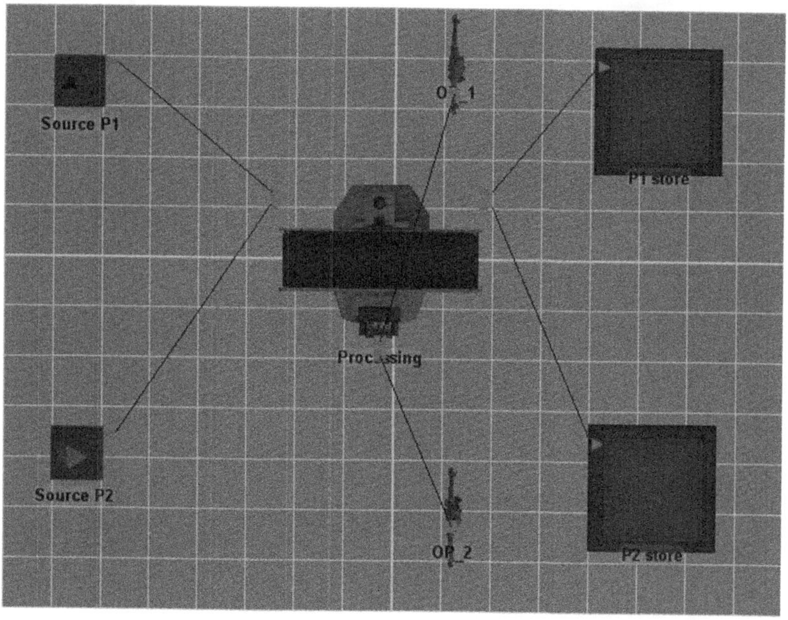

Figure 4.45. *The model with two sources, a processor, two queues and two operators. For a color version of this figure, see www.iste.co.uk/reveillac/modeling3.zip*

4.5.5.2. *Configuring the objects*

Under the TRIGGERS tab for the 'P1 source', add a function in the ONEXIT field to name the item and distinguish it by color (red in this case):

– Under SET LABEL, input OBJECT: Item; LABEL: "Product_1" and VALUE: 1;

– SET OBJECT COLOR with OBJECT: Item; COLOR: '`colorred(object)`'.

Under the TRIGGERS tab for the 'P2 source', add a function in the ONEXIT field to name the item and distinguish it by color (yellow in this case):

– Under SET LABEL, input OBJECT: item; LABEL: "Product_2" and VALUE: 1;

– Under SET OBJECT COLOR, input OBJECT: Item; COLOR: '`color yellow(object)`'.

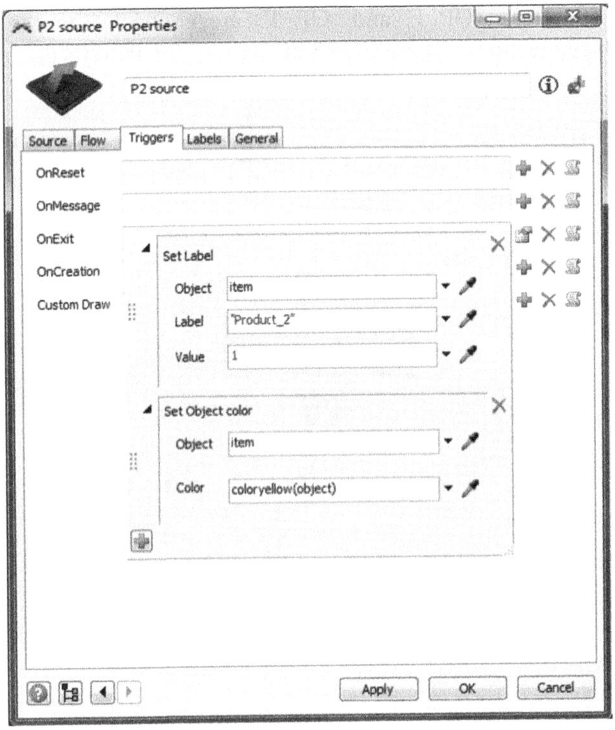

Figure 4.46. *The FLOW tab parameters for 'P2 source'*

Under the SOURCE tab for ‘Product_2’, input 2 in the ITEM TYPE field, to give the articles different types as well as names.

The items from the ‘Product_1’ source will be type 1 by default.

Defining types 1 and 2 will be useful for configuring the ‘Processing’ processor.

Under the FLOW tab of the ‘Processing’ processor:

– In the SEND TO PORT field: select CONDITIONAL PORT to set the processor to direct items according to their type – if the item equals ‘product_1’ then it goes to output port 1, if not it goes to output port 2. For CONDITION input: ‘`getlabel(Item, "Produit_1")`’; for TRUE input: 1 and for FALSE input: 2.

– Check the USE TRANSPORT option to use the operators and select MULTIPLE TRANSPORT OPTIONS with CASE 1, CENTERPORT: 1 and CASE 2, CENTERPORT: 2. Type 1 items will use operator ‘OP_1’ and type 2 items will use operator ‘OP_2’.

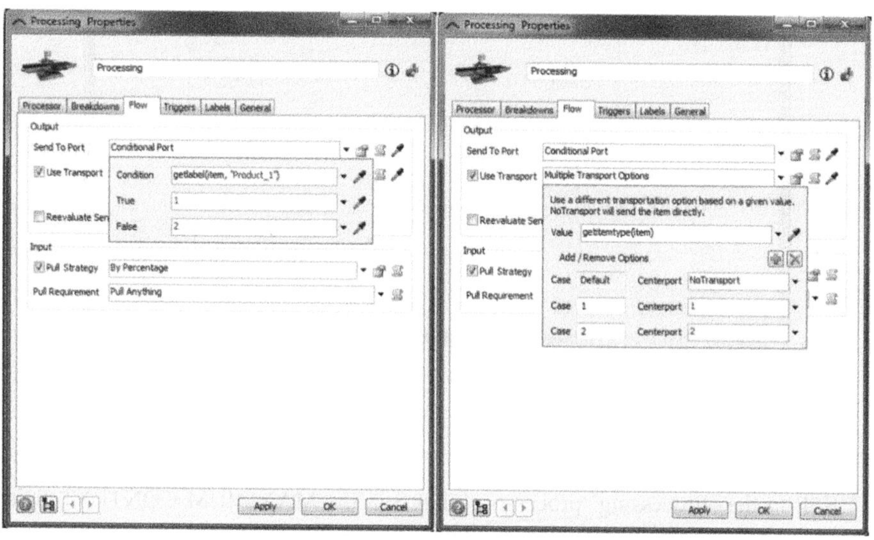

Figure 4.47. *The parameters for the OUTPUT zone under the FLOW tab for the ‘Processing’ processor*

– In the INPUT zone, check the PULL STRATEGY option and select BY PERCENTAGE to define the percentages of products to process, here 60% for input port 1 and 40% for input port 2.

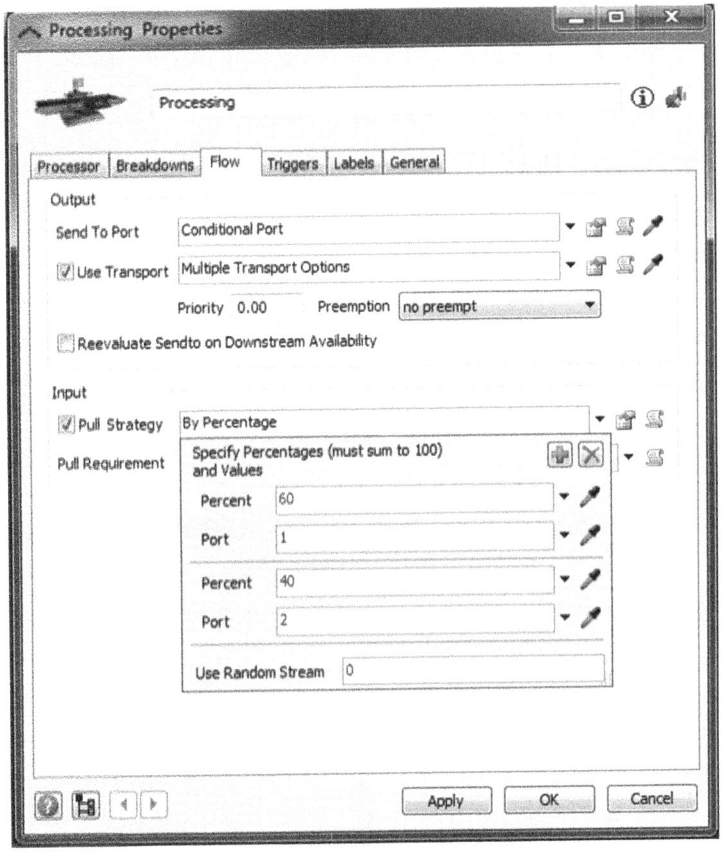

Figure 4.48. *The INPUT zone parameters under the FLOW tab for the 'Processing' processor'*

Still in the 'Processing' processor, input 5 in the MAXIMUM CONTENT field under the PROCESSOR tab to successively process five items.

4.5.5.3. *Testing the model*

Run the simulation after setting a STOP TIME (SPECIFY STOP TIME under RUN TIME in the auxiliary toolbar) equal to 2,400 s (40 min) so that a sufficient number of items are processed.

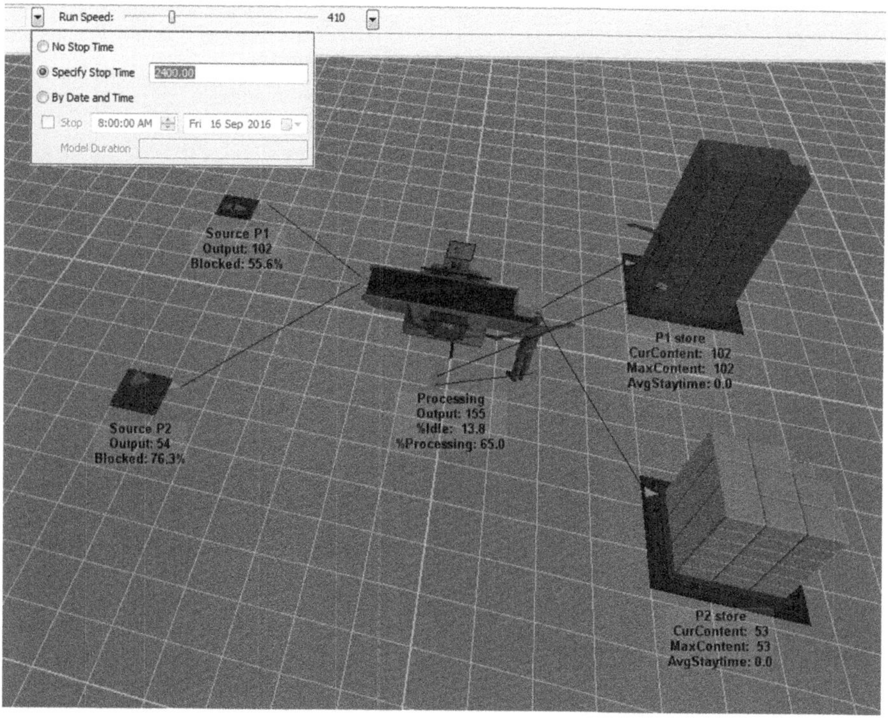

Figure 4.49. *The model during simulation, with a duration limited to 2,400 s. For a color version of this figure, see www.iste.co.uk/reveillac/modeling3.zip*

Each of the sources supplies items of different colors, red for 'P1 source' and yellow for 'P2 source'.

The 'Processing' processor processes up to five items in a row.

The two operators will collect all of the items of the color they have been allocated, i.e. red for 'OP_1' and yellow for 'OP_2'.

The stores individually refill themselves with items of a single color.

The name of the items, red or yellow, can be checked by double-clicking on them and looking under the LABEL tab.

4.5.5.4. *Add a dashboard*

A dashboard can be easily added to this model to display the store contents and the mean hourly production of each item (Figure 4.50).

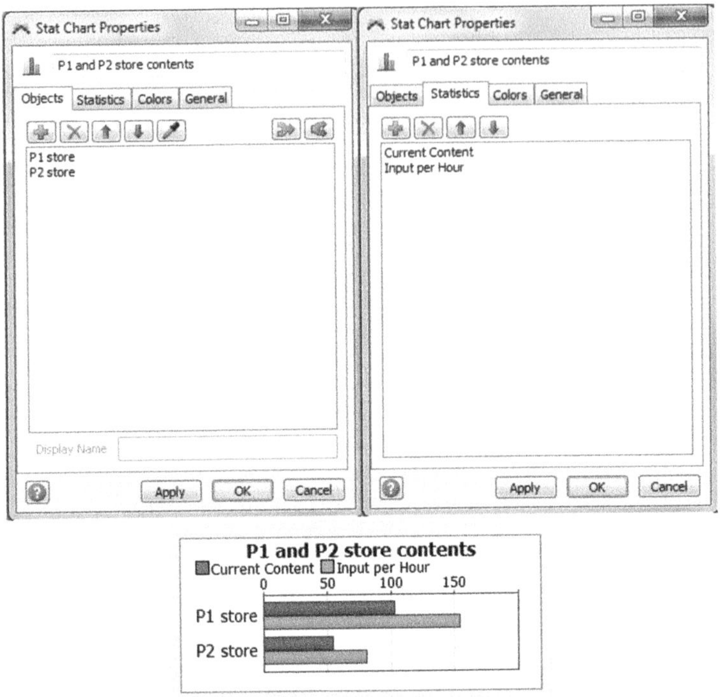

Figure 4.50. *The dashboard properties (above, OBJECTS and STATISTICS tabs) and the results obtained at the end of the simulation (below). For a color version of this figure, see www.iste.co.uk/reveillac/modeling3.zip*

4.5.6. *Timetables, groups and resources*

In many cases, it is often useful to be able to manage the timetables that are assigned to certain resources.

These resources can share common timetables so grouping them can be helpful.

In the example that follows, we will put these principles into action.

A source supplies a warehouse zone with products following a normal distribution: mean = 150 s and standard deviation = 50 s.

Four forklift truck operators, split into two teams of two people, then transport them in turns to a processing unit.

This unit carries out four operations on each product, with the following durations:

– operation 1: 90 s;

– operation 2: 30 s;

– operation 3: 50 s;

– operation 4: 35 s.

Operation 1 requires the presence of a technician, whereas the others are completely automated.

The two teams of workers observe the following working hours:

– team 1, composed of two forklift operators: 8:00 am–12:00 pm (morning);

– team 2, composed of two forklift operators: 1:00 pm–5:00 pm (afternoon).

The technician and the processing unit work from 8:00 am to 12:00 pm and from 1:00 to 5:00 pm.

Once the products have been processed, a robotic arm picks them up and places them in a storage zone.

In order to track the progress of the products through the model more easily, the products will be yellow boxes up to the processing unit, and green cylinders after leaving it.

The simulation will take place over 1 day, from 8:00 am to 5:00 pm.

4.5.6.1. *Model and connections*

Create the model shown in Figure 4.51, with the objects:

– one SOURCE: 'Products';

– two QUEUEs: 'Stockpile' and 'Storage';

– one DISPATCHER: 'Distrib';

– four forklift operators (TRANSPORTER): 'FL_1' to 'FL_4';

– one OPERATOR: 'Technician';

– one MULTIPROCESSOR: 'Processing';

– one ROBOT: 'Robot'.

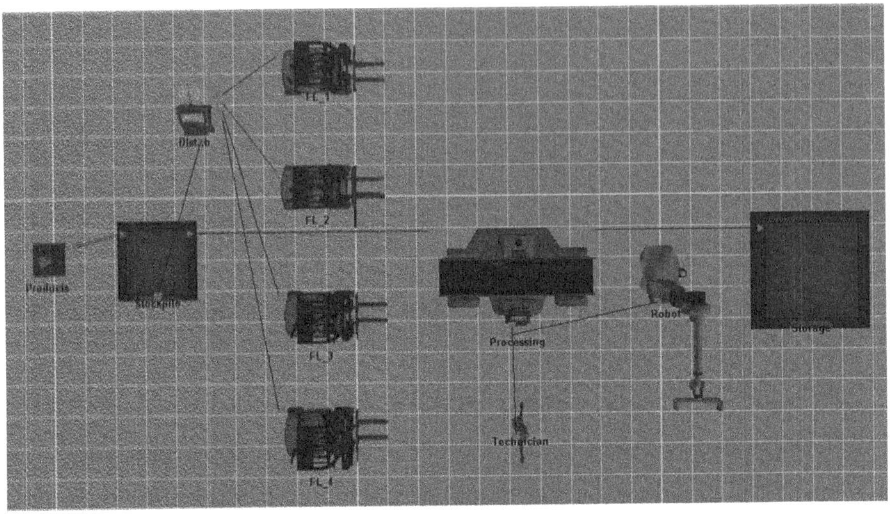

Figure 4.51. *The model and its connections. For a color version of this figure, see www.iste.co.uk/reveillac/modeling3.zip*

All of the fixed resources are connected to each other with connections of type CONNECT OBJECTS (a).

The distributor, 'Distrib', is connected to the four task executers, 'FL_1' to 'FL_4', with connections of type CONNECT OBJECTS (a).

The distributor, 'Distrib', is connected to the 'Stockpile' queue with a connection of type CONNECT CENTER PORTS (s).

The 'Technician' operator and the 'Robot' robot are connected to the 'Processing' multiprocessor with connections of type CONNECT CENTER PORTS (s).

4.5.6.2. *Configuring the objects*

We start by specifying the model configurations.

In the FILE menu, select MODEL SETTINGS and specify 8:00 am as the MODEL START TIME, then HH:mm:ss as the TIME FORMAT.

In the auxiliary toolbar, by RUN TIME, check BY DATE AND TIME, check STOP and set 17:00:00 as the stop time.

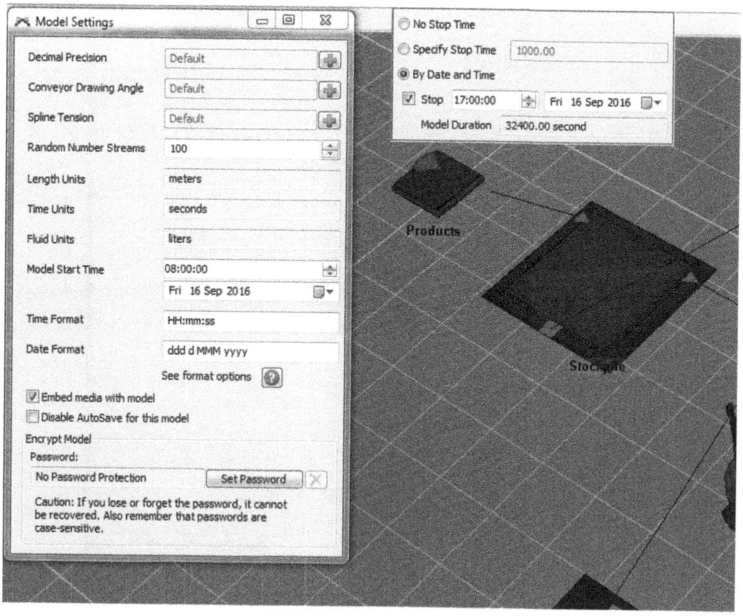

Figure 4.52. *Time settings for the model simulation*

We next define the timetables we need.

Go to the TOOLBOX tab (or the VIEW menu, then TOOLBOX).

Click on the add icon (green +) at top-left and select TIME TABLE.

In the window that opens, input 'Timetable_AM' in the NAME field.

In the TABLE tab, select DAILY REPEAT as MODE, then delete the contents of the cells from 12 PM to 5 PM to obtain the table shown in Figure 4.53, which represents the timetable for the morning team.

Repeat the operation a second time by creating a 'Timetable_PM' timetable with a timetable going from 1 PM to 5 PM. This represents the timetable for the team who work in the afternoons.

Finally, repeat the operation a third time by creating a 'Processing_unit' timetable, from 8 AM to 12 PM and from 1 PM to 5 PM. This is the timetable for the 'Processing' machine and its operator, 'Technician'.

These timetables should now be in the TOOLBOX navigation tree (Figure 4.54).

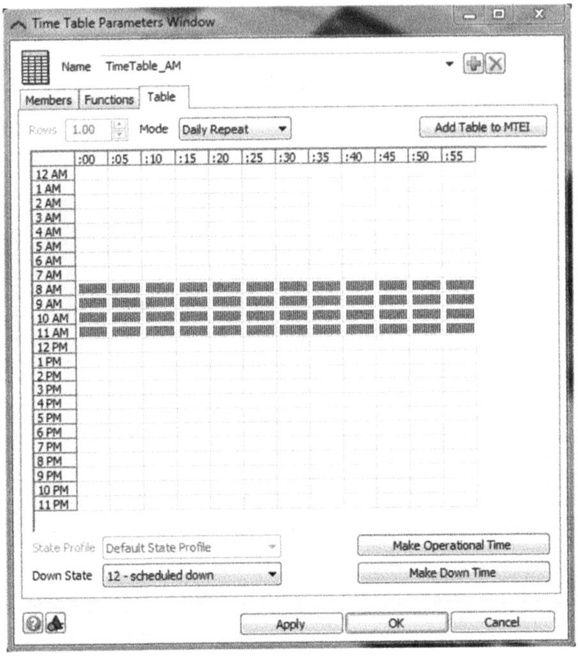

Figure 4.53. *The 'Timetable_AM' timetable*

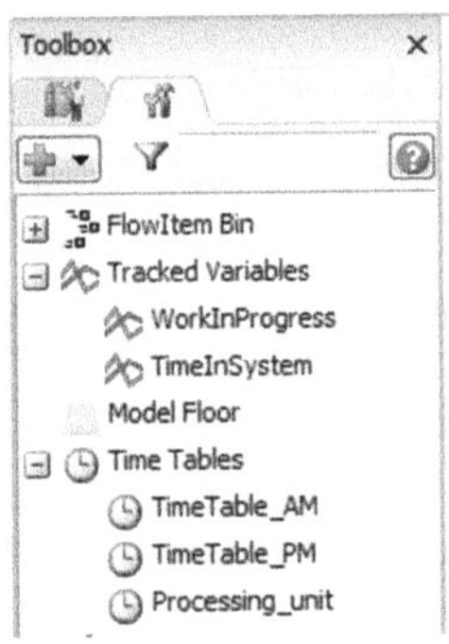

Figure 4.54. *The navigation tree in the TOOLBOX window with the three TIME TABLES: 'Timetable_AM'; 'Timetable_PM' and 'Processing_unit'*

We will now create two teams of workers.

Go to the VIEW menu and select GROUPS. The OBJECT GROUPS window appears on the right-hand side of FlexSIM.

Click on the top-left icon, CREATE A NEW GROUP and change the name of the group to 'AM_Team'.

Click on the pipette to select it, then click on forklift truck operator, 'FL_1' in the model to add them to the list of members. Do the same for 'FL_2'.

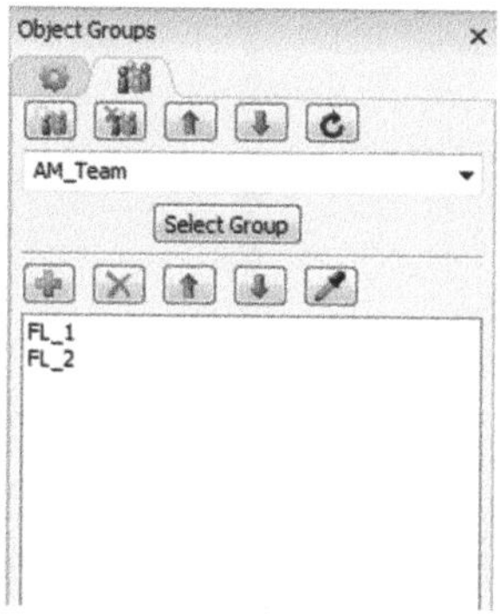

Figure 4.55. *The OBJECT GROUPS window for the 'AM_Team', containing 'FL_1' and 'FL_2' as members*

In the same way, create a second group, 'PM_Team' and add forklift truck operators, 'FL_3' and 'FL_4' to it.

We must now assign the calendars created previously to each of the objects, 'FL_1' to 'FL_4', 'Technician' and 'Processing'.

Reopen the three timetables by double-clicking on them in the TOOLBOX navigation tree, then go to the MEMBERS tab and click on the + icon to add the resources using these timetables.

The members can be added in two ways:

– selecting objects from the different classes (BROWSE BY LIBRARY CLASS);

– selecting groups of objects (BROWSE BY GROUP);

Using the two icons present in the dialog box (Figure 4.56).

The timetables should have the following members:

– 'Timetable_AM': 'AM_Team', i.e. 'FL_1' and 'FL_2';

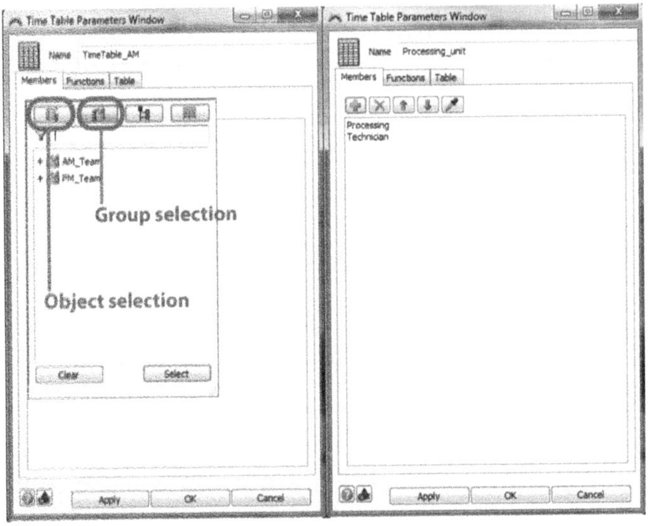

Figure 4.56. *Selecting the 'AM_Team' group, comprising 'FL_1' and 'FL_2', for the 'Timetable_AM' timetable*

– 'Timetable_PM': 'PM_Team', i.e. 'FL_3' and 'FL_4';

– 'Processing_unit': 'Processing' multi-processor and 'Technician' operator.

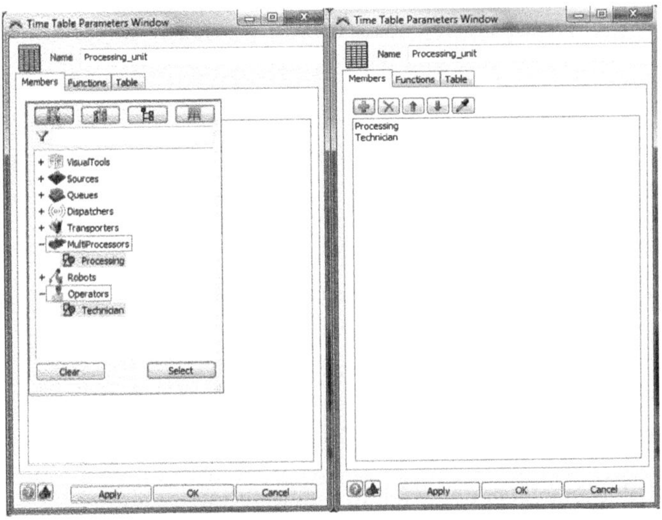

Figure 4.57. *Selecting the 'Processing' and 'Technician' objects for the 'Processing_unit' timetable*

Under the SOURCE tab for the 'Products' source, select a normal distribution in the INTER-ARRIVALTIME field, with the following parameters: '`normal(150, 50, 0)`'.

Under the TRIGGERS tab, select SET OBJECT COLOR in the ONEXIT field and set 'item' for OBJECT and 'yellow' for COLOR, i.e.: '`coloryellow(object)`'.

Open the 'Stockpile' queue, go to the FLOW tab and check USE TRANSPORT so that it can use task executers related to the distributor, 'Distrib'.

Open the 'Processing' multi-processor. Under the MULTIPROCESSOR tab, add (+ icon) the four operations, 'Operation_1' to 'Operation_4' with their respective durations (90, 30, 50 and 35 s) in the TIME PROCESS field.

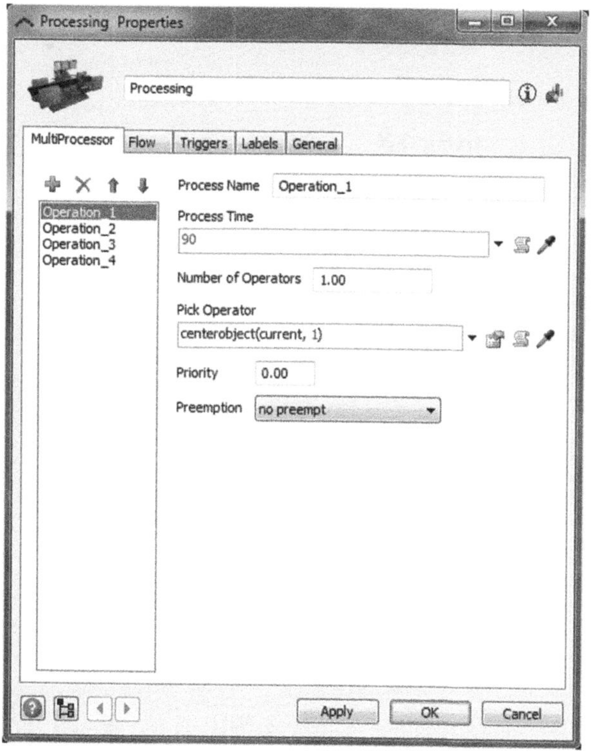

Figure 4.58. *Configuration of the MULTIPROCESSOR tab for the 'processing' object. Here, 'Operation_1' is visible, with a PROCESS TIME of 90 s and using one operator (NUMBER OF OPERATORS) connected to central port 1 (PICK OPERATOR: 'center object (current, 1)')*

For ‘Operation_1’, input 1 in the NUMBER OF OPERATORS field. This indicates that one operator will be used to carry out this operation. By default, this operator is used *via* central port 1 (PICK OPERATOR: ‘`centerobject(current, 1)`’.

Under the FLOW tab, check the USE TRANSPORT option and set the port that will be used for the ‘Robot’ task executer: ‘`centerobject(current, 2)`’.

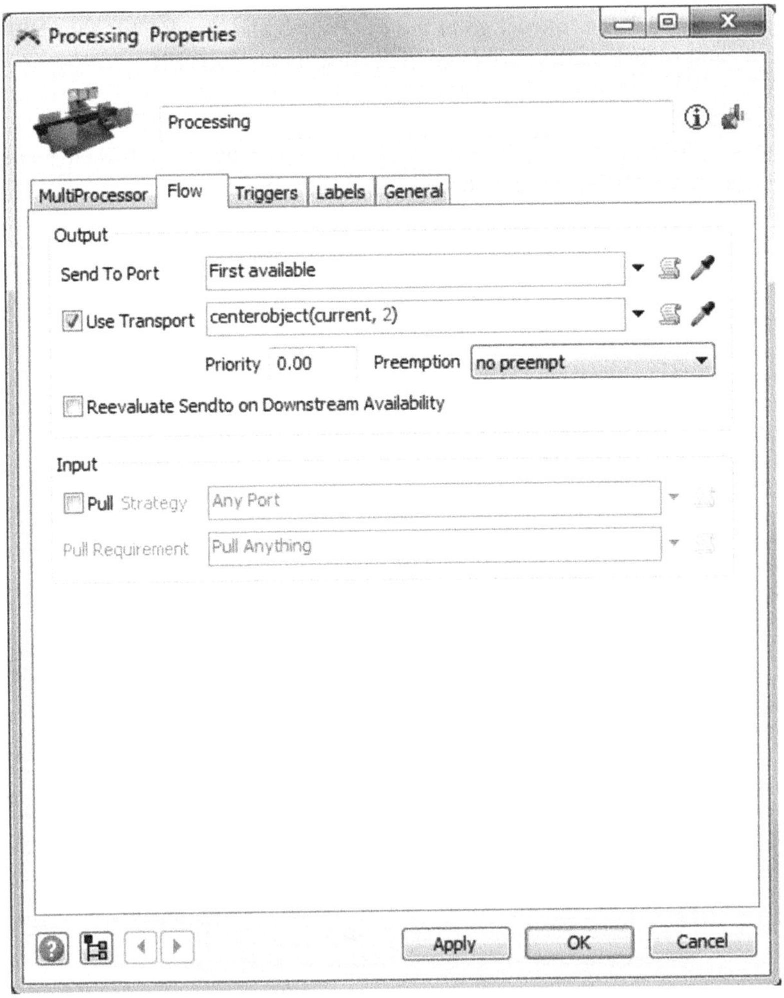

Figure 4.59. *Configuration of the FLOW tab for ‘Processing’*

Under the TRIGGERS tab, for the ONEXIT field, in VISUAL, select SHAPE: 2 – CYLINDER.3DS under CHANGE 3D SHAPE. Also add SET OBJECT COLOR with the COLOR: '`colorlime(object)`'.

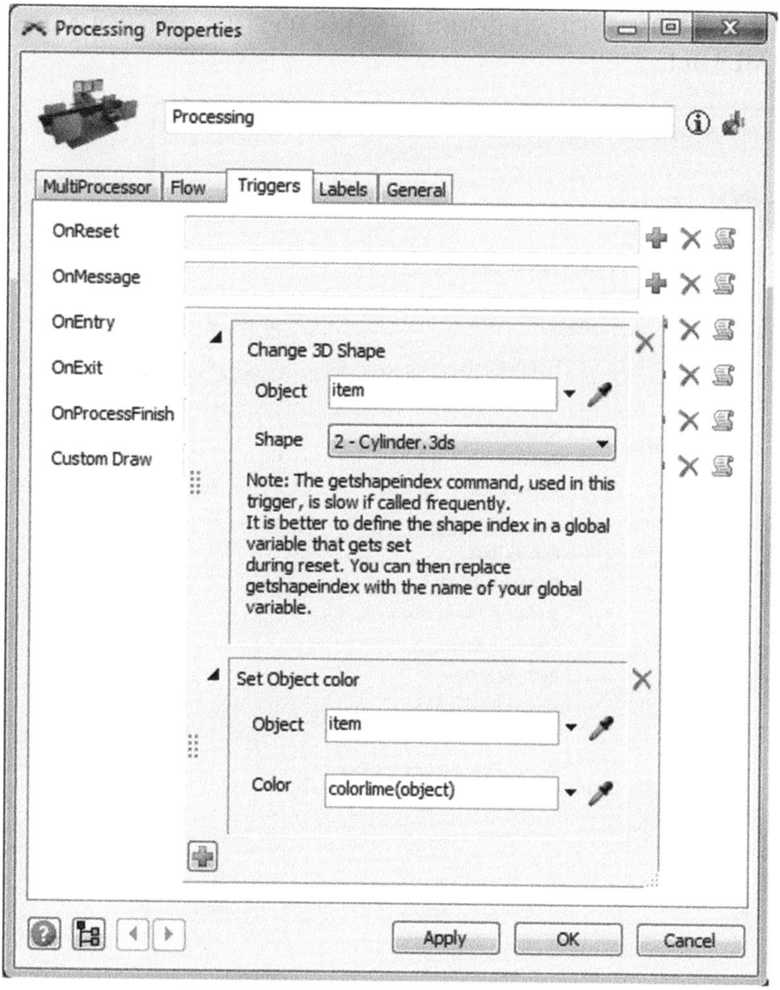

Figure 4.60. *Configuring the TRIGGERS tab for the 'processing' object*

In order to satisfy the problem statement (see section 4.5.6) and make each of the two workers in each team work in turn, a ROUND ROBIN IF AVAILABLE option for the distributor's properties is needed, but this does not exist.

We will hence copy its code from another object, for example the 'Processing' multi-processor, then paste it and modify it.

Open the properties window of the 'Processing' multi-processor, go to the FLOW tab and select ROUND ROBIN in SEND TO PORT, then ROUND ROBIN IF AVAILABLE.

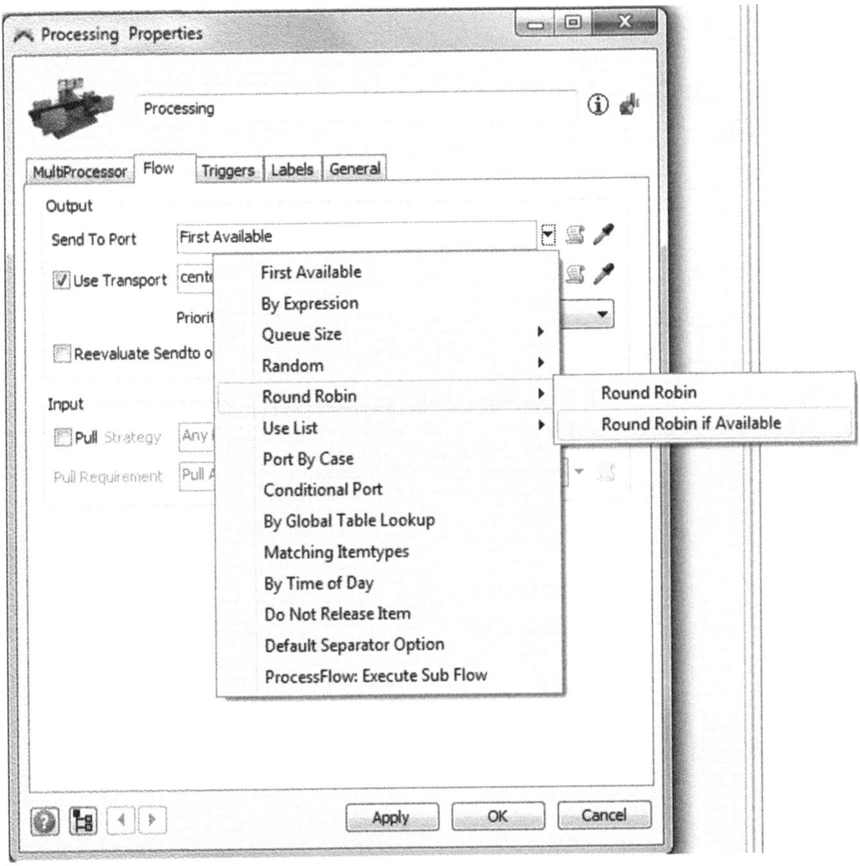

Figure 4.61. *Configuring the FLOW tab of the 'Processing' object, with SENDTOPORT set to ROUND ROBIN IF AVAILABLE*

Click on the icon that looks like parchment to see the code. Select from '`treenode item = param1);`' to '`return returnvalue;`' (lines 1 to 31) then right-click and select COPY (or use CTRL C).

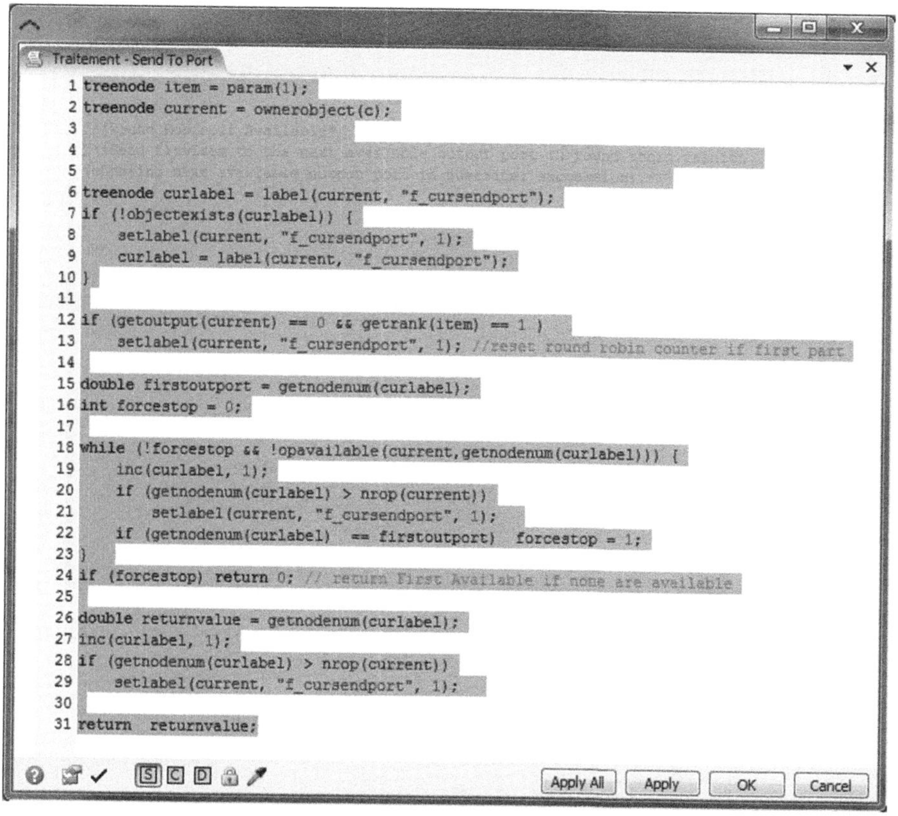

Figure 4.62. *The ROUND ROBIN IF AVAILABLE code selected and ready for copying*

Click on CANCEL to close the code window, then on CANCEL to close the multi-processor window without saving any changes.

Open the 'Distrib' distributor properties, and then under the DISPATCHER tab, opposite the PASS TO field, click on the parchment icon, select the existing code and replace it by right-clicking and selecting PASTE (or use CTRL V).

Select and delete lines 12 and 13, which are not needed in this case.

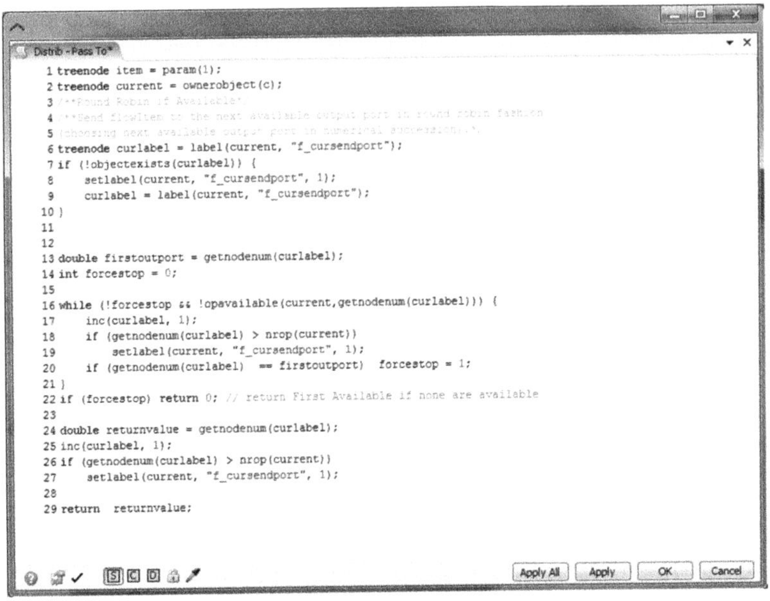

Figure 4.63. *The pasted and modified code for PASS TO in the 'Distrib' distributor*

Figure 4.64. *The DISPATCHER tab of the 'Distrib' properties window, after pasting and modifying the code*

Confirm by clicking on OK. The ROUND ROBIN IF AVAILABLE option should be present in the PASS TO field. Click on OK to confirm and close the 'Distrib' properties window.

The 'Robot', 'Storage' and four forklift truck operator objects, 'FL_1' to 'FL_4', keep their default parameters.

4.5.6.3. *Testing the model*

At 8:00 am, when the simulation starts, the storage zone fills up and the forklift truck operators, 'FL_1' and 'FL_2, collect the products (in the form of yellow boxes) in turn, to take them to the processing unit. Operator 'OP_ 1' takes position in front of the processing unit ready to carry out his work once the first operation is complete.

At the end of this unit, a robot picks up the processed products, which are now green cylinders, and places them in the storage zone.

Between 12:00 and 1:00 pm, the forklift truck operators and processing unit do not work.

Work starts up again after 1:00 pm, now with the forklift truck operators, 'FL_3' and 'FL_4', who transport the products in turn.

At 5:00 pm, the simulation ends.

At 12:00 pm, when the forklift truck operators, 'FL_1' and 'FL_2', stop transporting the products, the animation of the products that have not been delivered to the processing unit, and are still being transported, freezes. The same happens when the forklift truck operators, 'FL_3' and 'FL_4', stop working at the end of the day, at 5:00 pm.

To avoid this problem, we can force the products en route to the processing unit to be delivered when the timetable finishes and commands the 'Distrib' distributor to stop giving products to the forklift truck operators at 12:00 and 5:00 pm exactly, by closing their ports.

Go to the TOOLBOX navigation tree and open the 'Timetable_AM' TIME TABLE assigned to the forklift truck operators, 'FL_1' and 'FL_2'. Under the FUNCTIONS tab, select STOP INPUT in the DOWN FUNCTION field and RESUME INPUT in the RESUME FUNCTION field.

Configure the 'Timetable_PM' timetable for the forklift truck operators, 'FL_3' and 'FL_4', in the same way.

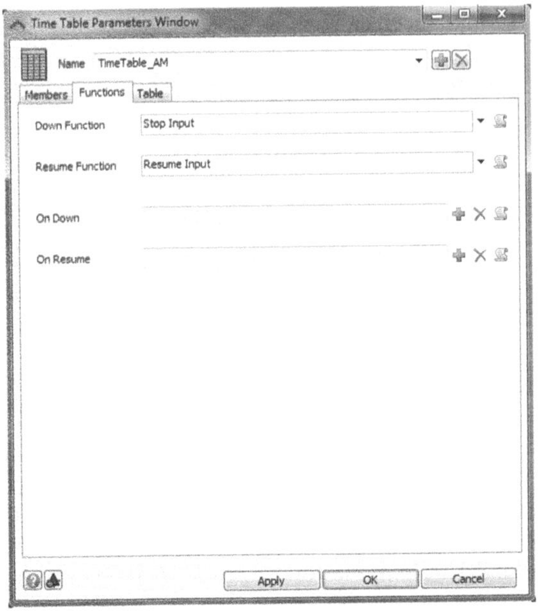

Figure 4.65. *The FUNCTIONS tab configuration for the 'Timetable_AM' timetable used to stop the forklift truck operators working at the end of their shift (12:00 pm)*

When a port is closed, a red bar appears in front of the resource's green arrow.

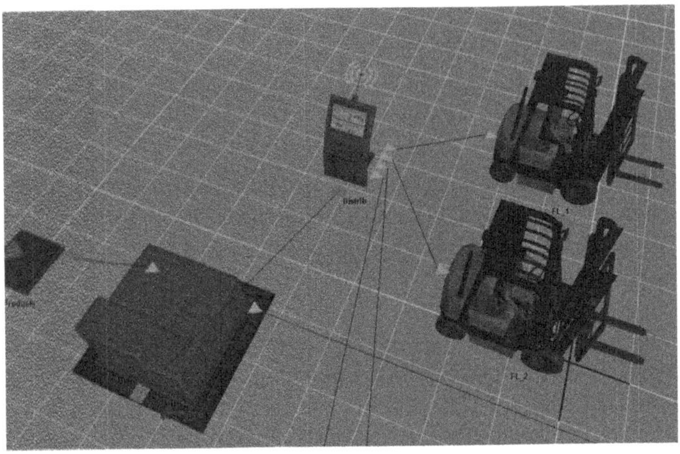

Figure 4.66. *The two forklift truck operators 'FL_1' and 'FL_2' (on the right) with closed input ports. A red bar can be seen in front of their green arrows. For a color version of this figure, see www.iste.co.uk/reveillac/modeling3.zip*

4.6. Experimenter

Many models are based upon statistical distribution functions, such as the normal, exponential or beta distributions, and random phenomena that supply different values each time a simulation is run.

The simulation must be run a sufficient number of times for the results to be reliable.

FlexSim has a tool specifically for this purpose, called the Experimenter.

4.6.1. *Constructing the model*

To see how the Experimenter works and what it does, construct the very simple model shown in Figure 4.67.

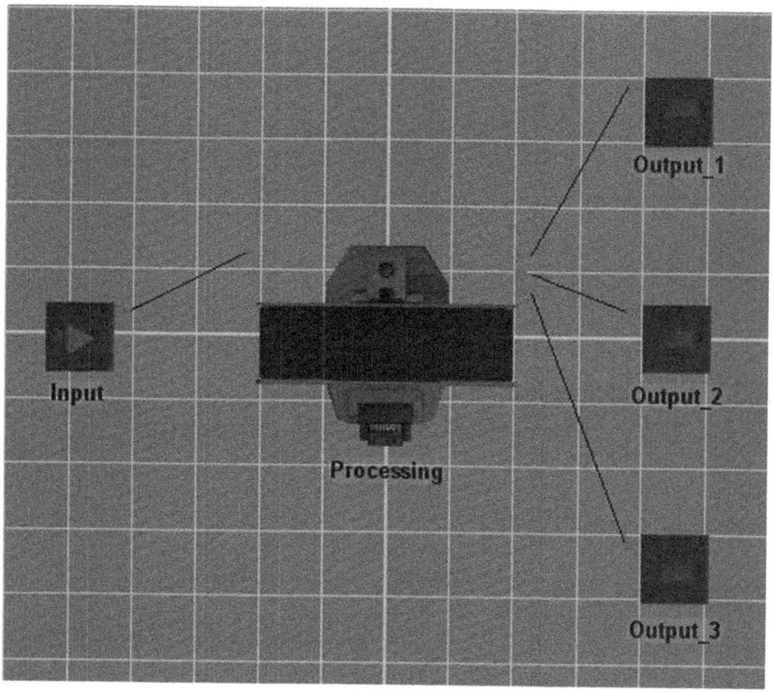

Figure 4.67. *A simple model for testing the Experimenter. For a color version of this figure, see www.iste.co.uk/reveillac/modeling3.zip*

It is made up of five objects:

Object	Number	Name	Parameters
SOURCE	1	'Input'	SOURCE tab: INTER-ARRIVALTIME - 'normal(100, 10, 0)' TRIGGERS tab: ONEXIT – SET ITEMTYPE AND COLOR – ITEM TYPE – 'duniform(1,3)'
PROCESSOR	2	'Processing'	PROCESSOR tab: PROCESS TIME – VALUES BY CASE CASE 1: 'exponential(0.0, 100, 0)' CASE 2: 'exponential(0.0, 50, 0)' CASE 3: 'exponential(0.0, 10, 0)' FLOW tab: SEND TO PORT – PORT BY CASE – CASE 1, PORT 1; CASE 2, PORT 2; CASE 3, PORT 3
SINK	3	'Output_1' 'Output_2' 'Output_3'	

One source supplies three different items to one processor. The processing time and distribution to the three outputs are different for each item type.

4.6.2. *Adding the dashboard*

The Experimenter draws its results from one or more dashboards.

A dashboard must therefore be added to the model using the DASHBOARD icon from the main toolbar and the ADD A DASHBOARD option.

Next, select an AVERAGE CONTENT type dashboard from the LIBRARY window, under the CONTENT heading.

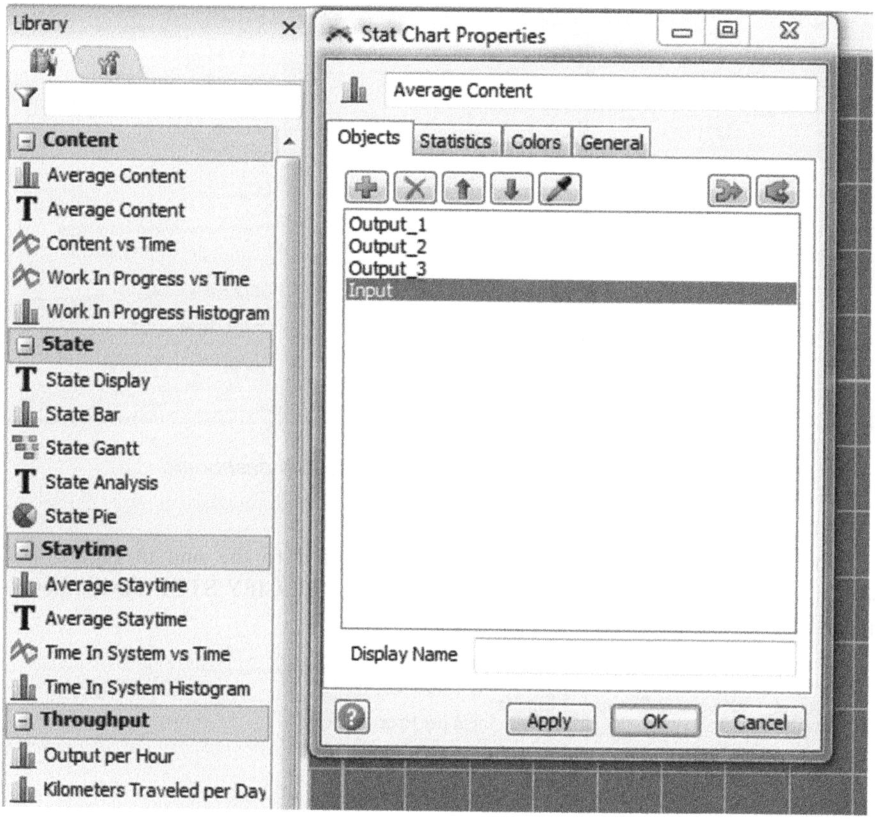

Figure 4.68. *The dashboard LIBRARY window (left) and the AVERAGE CONTENT properties window*

Under the OBJECTS tab for this dashboard's properties, add the objects, 'Output_1', 'Output_2', 'Output_3' and 'Input'.

Under the STATISTICS tab, add TOTAL INPUT, TOTAL OUTPUT, OUTPUT PER HOUR and INPUT PER HOUR.

Under the GENERAL tab, select LINE CHART for DISPLAY TYPE.

Name the dashboard, 'Inputs-Outputs'.

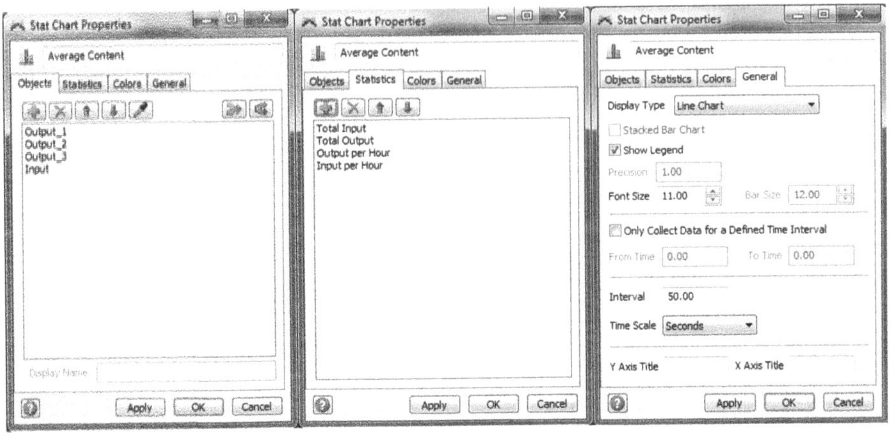

Figure 4.69. *The configuration tabs for the dashboard*

Upon running the simulation, a dashboard similar to the one in Figure 4.70 should appear. It shows the results obtained after 1 h (SPECIFY STOP TIME: 3,600 s under RUN TIME in the auxiliary toolbar).

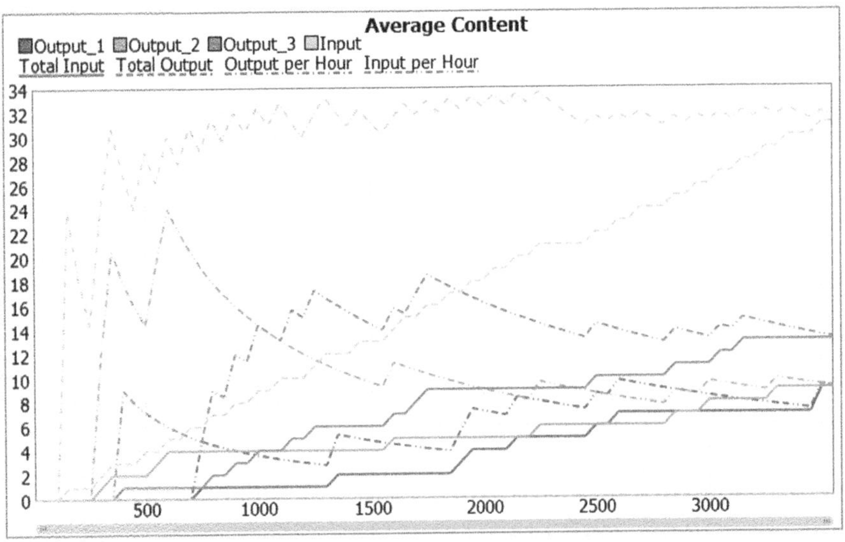

Figure 4.70. *The dashboard for the model at the end of the simulation. For a color version of this figure, see www.iste.co.uk/reveillac/modeling3.zip*

4.6.3. *Configuring the Experimenter*

Go to the STATISTICS menu and select EXPERIMENTER…

The SIMULATION EXPERIMENT CONTROL window should open. Select the PERFORMANCE MEASURES tab and, using the plus sign icon, add:

– Inputs/Outputs – TOTAL INPUT – 'Output_1';

– Inputs/Outputs – TOTAL INPUT – 'Output_2';

– Inputs/Outputs – TOTAL INPUT – 'Output_3';

– Inputs/Outputs – TOTAL OUTPUT – 'Input';

– Inputs/Outputs – OUTPUT PER HOUR – 'Input'.

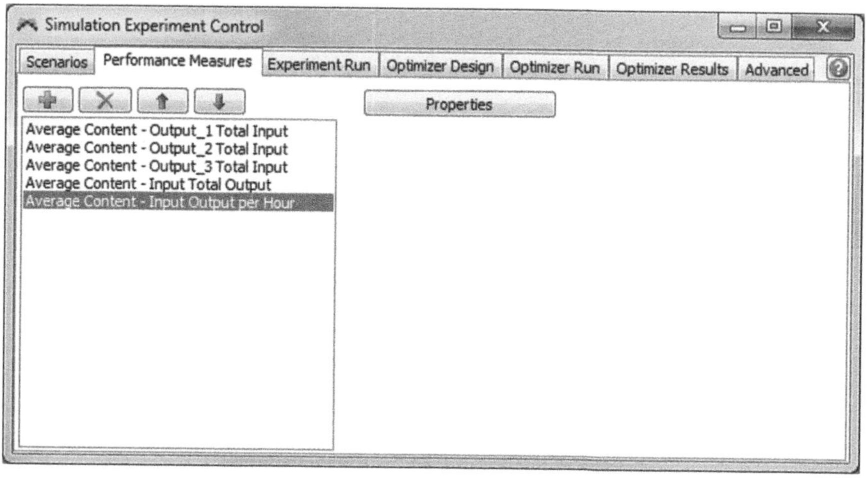

Figure 4.71. *The PERFORMANCE MEASURES tab in the Experimenter window*

Click on the EXPERIMENT RUN tab and input 100 in the REPLICATIONS PER SCENARIO field, and 3,600 s in the RUN TIME field.

The Experimenter will carry out the scenario 100 times for a simulation duration of 1 h (3,600 s).

Click on RUN EXPERIMENT to run the Experimenter. A progress bar called CURRENT SCENARIO will show the progress, going from red to green.

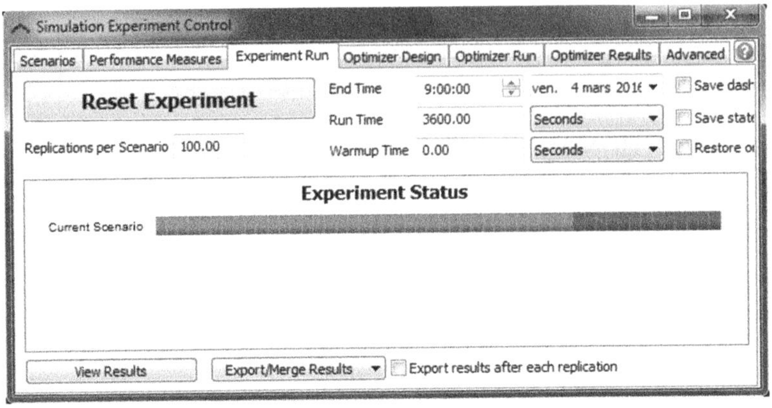

Figure 4.72. *The progress bar in the EXPERIMENT RUN tab during operation. For a color version of this figure, see www.iste.co.uk/reveillac/modeling3.zip*

Once the Experimenter has simulated the full number of scenarios, a RESET EXPERIMENT button appears. The PERFORMANCE MEASURE RESULTS window summarizing all of the results can be accessed by clicking on VIEW RESULTS.

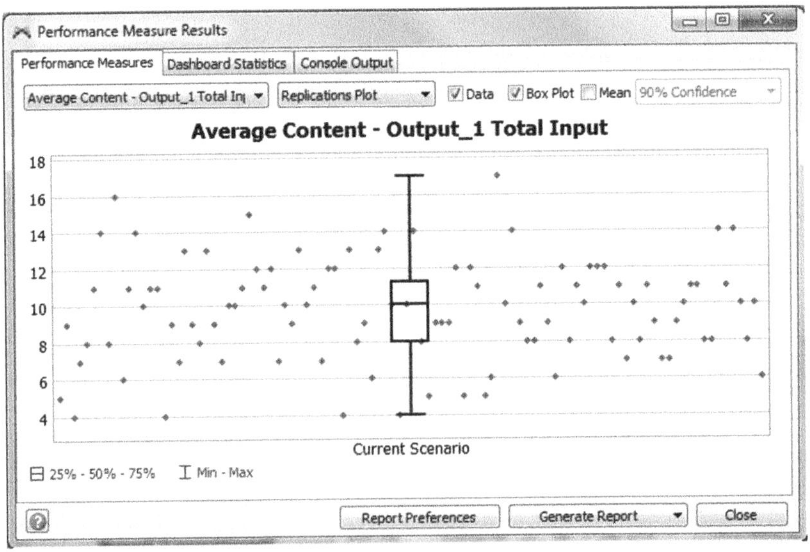

Figure 4.73. *The results summary window*

A wide range of different graphics relating to the parameters chosen in the dashboard can be displayed using the PERFORMANCE MEASURES subtabs.

A dashboard for each of the 100 scenarios can also be displayed using the DASHBOARD STATISTICS tab provided that the SAVE DASHBOARD DATA FOR EACH REPLICATION option is checked in the EXPERIMENT RUN tab of the SIMULATION EXPERIMENT CONTROL window (Figure 4.72) before running the Experimenter.

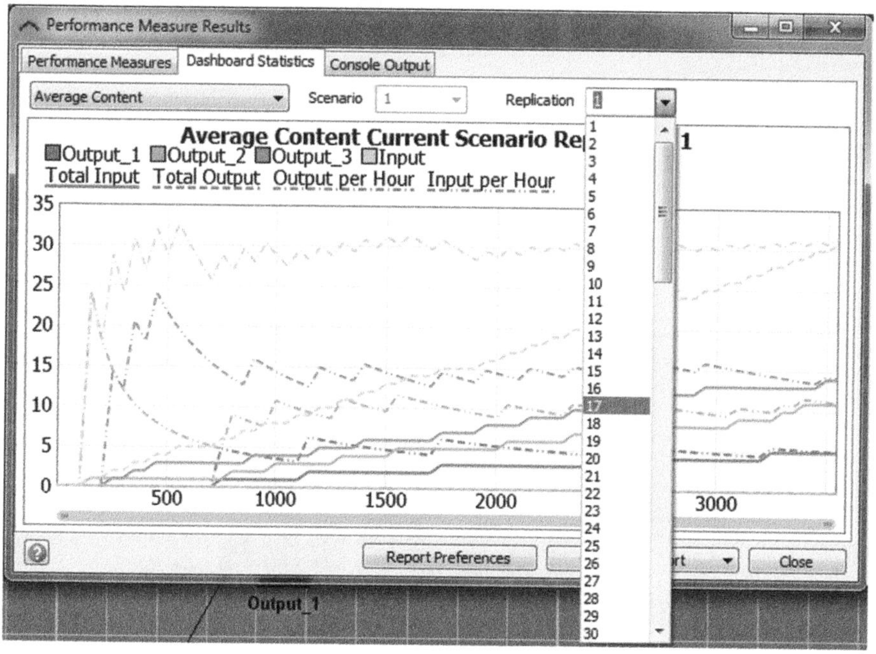

Figure 4.74. *Accessing the 100 dashboards for each of the scenarios run by the Experimenter. For a color version of this figure, see www.iste.co.uk/reveillac/modeling3.zip*

At the bottom of the PERFORMANCE MEASURE RESULTS window, there is a GENERATE REPORT button. This creates a report summarizing all of the results in the form of a web page, with two possible formats:

– IN REPORT FORMAT: a page with the results placed one after each other (Figure 4.75).

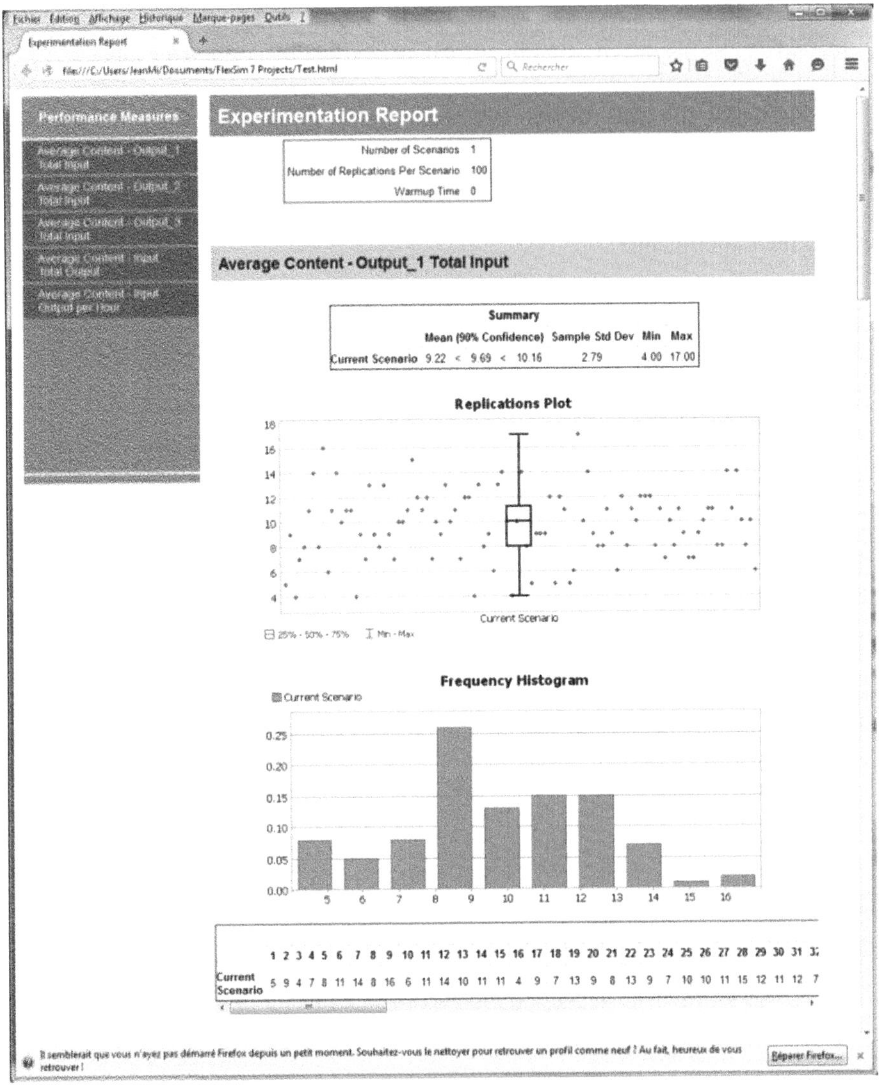

Figure 4.75. *An IN REPORT FORMAT type report*

– IN WEB VIEWER FORMAT: an interactive page with an interactive selection of results that can be configured using the drop-down lists and buttons (Figure 4.76).

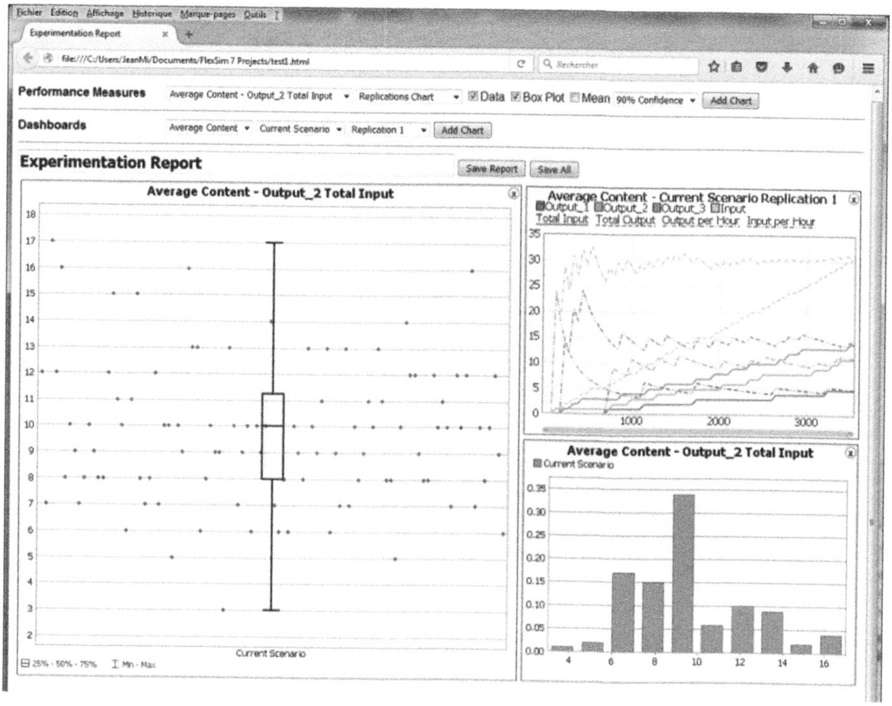

Figure 4.76. *An IN WEB VIEWER FORMAT type report*

4.7. Concluding remarks

In slightly over 80 pages, we have only scratched the surface of the different options provided by the FlexSim software for managing warehouse storage. Its power and capabilities are extensive, and most logistics simulations, across any field of work, can be modeled using this software.

Some examples of capabilities not discussed here include exchanging data with Excel, importing 2D or 3D objects, writing scripts, creating navigation interfaces, managing flow graphics, animating objects and implementing guided tours.

For more depth, the reader is referred to the software's user manual (in English), which is directly accessible from the start-up screen *via* the USER MANUAL icon. It contains a section dedicated to tutorials delivered as collections of multiple lessons.

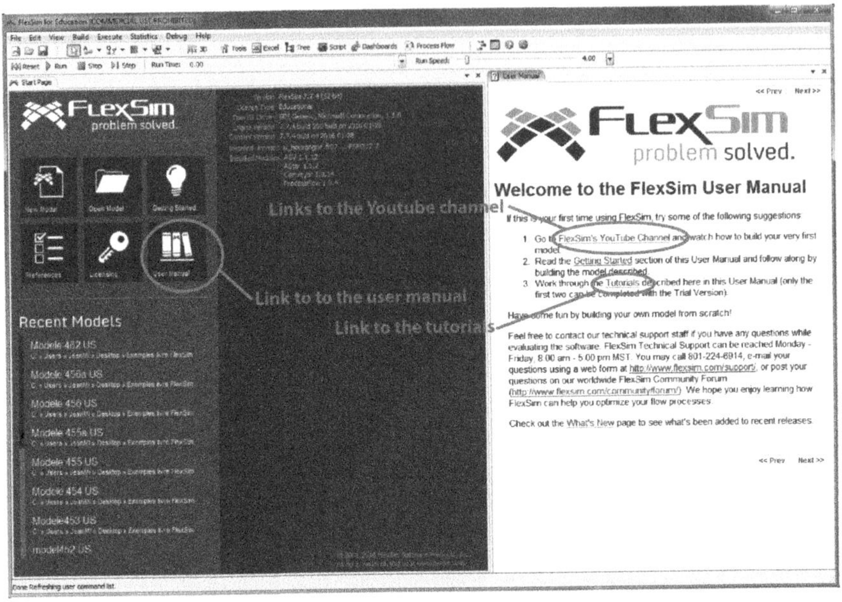

Figure 4.77. *Access to the user manual, with links to the tutorials and YouTube videos*

There is also a YouTube channel dedicated to FlexSim, where a large number of videos are available:

http://www.youtube.com/user/FlexsimSoftware?feature=watch

You are also encouraged to consult the publisher's Website: www.flexsim.com.

Conclusion

This third volume has presented a range of software, together with examples and exercises mostly drawn from real-world situations. Although it is not possible to cover all of the logistics problems that can arise from day to day, hopefully enough of the elements required for adapting or creating new solutions have been presented to enable you to deal with the majority of potential scenarios.

Innovation is relentless, new applications appear or are improved every day, and universities, laboratories and publishers continually turn to new sectors where optimization and logistics simulation are still in their infancy.

New approaches and techniques, influenced by the appearance of new technologies, only add to the ever-growing demand from industrial and institutional bodies.

Sustainable development, ecological concerns, lasting quality improvements and developments in norms and standards are leading to rapidly changing procedures. The new specifications that these procedural changes entail are ever more demanding to integrate and manage.

Product lifecycles change, their traceability improves, their recycling must be considered, their storage, distribution and transport migrate to new organizations, and all the while globalization makes competition fiercer and competitiveness essential.

As discussed in the introduction, the convergence of data flows, product flows and consumer behavior within immense, systematically recorded, categorized and organized databases has given birth to "big data".

These inexhaustible reserves of data are no longer on a human scale, and must be handled by high-performance computer applications, subject to regulations regarding transparency, compliancy and quality.

There are a wide range of applications dedicated to logistics. These range from the most traditional, such as enterprise resource planning (ERP), and the classic, such as spreadsheets, to more specialized and specific applications including flow managers and simulators.

Many of them work together and share databases or "data warehouses" in order to standardize formats and procedures while eliminating redundancy and repeats.

Today's computing power, combined with a degree of standardization in data formats (for example, XML), allows variables to be easily imported or exported, making them readily exploitable. Furthermore, software such as computerized flow simulators that require powerful computational resources can now be run on classic machines with any operating system.

Hardware and software are continually in development, and users always want more. Whereas software used to display sequences of data in text format, nowadays, their dashboards are as attractive as they have ever been, combining text, images and graphics. Logistics has entered the 3D and virtual domains and it is no longer enough to simply enter the warehouse specifications needed to render it in 2D. These days, logisticians want to be able to enter a warehouse and navigate around it to better understand, analyze and improve the processes implemented to meet the design brief.

Logisticians must also convince their superiors, decision makers, and investors, of the reasoning behind their decisions. A 3D animated model is optimal for conveying this reasoning.

In conclusion, logistics is progressing, even if the journey is often tortuous. It keeps moving forward and, at this early stage of the 21st C*entury*, it has foundations laid in all fields, even the most unlikely. Across industrial and institutional sectors, in any sphere of activity (manufacture, storage, hospital management, construction, architecture, security, inspections, defense, transport, land management, leisure and relaxation, data management and many others), logistics is everywhere and has continually growing influence. It has gained critical importance, and even become a societal issue given its capability to shape the world in response to the pressures exerted by the globalization of commerce and trade, diversity and the enormous quantities of data to be processed.

The society of tomorrow can only reach its full potential by mastering logistics, extending its scope toward opportunities that are emerging, such as the internet of things, or rarely considered, such as sustainable development. It may also need to simply become more "human".

Glossary

Affine function: a function in the form of $y = ax + b$. The variable a is called the leading coefficient and the variable b, y-intercept. A linear function is a straight line. When it goes through the origin, in this particular case (y-intercept null), we then get a linear function ($y = ax$).

Analytical resolution: in this type of problem solving, the approach to the solution is based on a mathematical procedure that provides precise and rapid result(s). In parallel with analytical resolution, we can choose a graphic solution. In linear programming, the simplex method is a type of analytical resolution.

Branch and bound: a method that consists of finding the optimum solution to a problem of combinatorial optimization. It is built around the concept of separation (branch) that carves the group of solutions into subsets of a smaller size and optimistic assessment in increasing (bound) these same subsets.

BRIC (Brazil, Russia, India and China): acronym designating the four principal developing countries.

Canonical form (or standard form): form of a problem in linear programming in which the constraints are expressed by inequations.

CAPM (computer-aided production management): software dedicated to managing all of the activities related to production in a business. Here, we can find, in the form of modules, the management of orders, the management of articles, the management of resources, the management of bills or materials and ranges of products, management of planning de fabrication, the management of purchases, etc.

Cardinality: a pair of values attached to a relation in the entity-relationship model (relational model). For a relationship (or association) between two entities, two pairs must be specified. The cardinalities express the number of times that the occurrence of an entity can be taken into account in the relationship, at minimum or maximum. There are four typical cardinality pairs: $(0, 1)$, $(1, 1)$, $(0, n)$ and $(1, n)$, n being the equivalent of several.

Concentration: also known as density. For a traffic model, this is the number of vehicles present on a given length of a route at a defined time.

Constraint: factor or condition that an optimization problem must satisfy.

CONTRAM: mesoscopic model of road traffic.

CPM (critical path method): method for calculating a critical path on a graph for planning tasks.

Decision-making aid: a group of techniques based on probability theory, graph theory or even operational research. They offer the opportunity for a human actor to opt for the best possible solution when faced with problems that are mainly industrial, financial or other.

Density: for road traffic, density is defined as the number of vehicles per unit length of the roadway.

Deterministic algorithm: an algorithm that carries out a predefined process in order to resolve a problem.

Dual: in linear programming, the initial problem, called "primal", is associated with another linear problem known as a "dual" problem. In some ways, the "dual" is the symmetrical to the "primal".

DynaMIT: mesoscopic model of road traffic.

Economic function (or cost function): function to be optimized in the context of linear programming.

Fastlane: mesoscopic model of road traffic.

Flow: for road traffic, flow is the number of vehicles passing a reference point per unit of time, vehicles per hour.

GRH (Gazis, Herman and Rothery): microscopic model of road traffic.

Heuristic: refers to a method that proceeds by successive approaches progressively eliminating the alternatives to keep only a limited group of solutions leaning toward optimality.

Hybrid: refers to a method of traffic simulation model combining microscopic and macroscopic modes with the possibility, among others, to go from one to the other via a simple change in scale.

ITS (Intelligent Transportation Systems): application of NICT designed for transportation.

Kirchhoff's Law: also known as the "nodal rule". This is a law that shows the conservation of energy in an electrical circuit. The sum of the intensity of the currents that enter into a node is equal to the sum of the intensities of the currents that come out of it. By extension, this law can be found in operational research, among other things, with the management of floats.

Linear function: see *affine function*.

LWR (Lighthill, Witham, Richards): macroscopic traffic model known as an equilibrium model.

Macroscopic: simulation mode within a traffic model. In this case, the flow approach of the network is global and involves large areas.

Maximization: refers to a problem of optimization when its objective must lead toward a maximum.

Mesoscopic: intermediate traffic simulation mode between microscopic and macroscopic.

Microscopic: simulation mode within a traffic model considering each of the vehicles and their interactions. This mode works on zones with small dimensions (crossings, crossroads, small groups of streets).

Minimization: refers to an optimization problem when its goal must must lead toward a minimum.

MPM (métra potential method): a method invented by Bernard Roy in 1958. It allows for the representation and optimization of a project in the form of a network-type diagram showing, among other things, the intertask dependencies while reducing delays and by constructing a critical path. With MPM, the tasks are symbolized by nodes (vertices) unlike the PERT method.

MST (minimum spanning tree): see *MWST*.

MWST (minimum weight spanning tree): a tree that connects all of the vertices of a graph and where sum of the weight of each one of its edges is minimal.

Normal distribution: also known as "Gaussian law" or "Gauss Laplace distribution". This is the law of probability that is most often used in statistics. Its curve is known as a "bell curve" due to its shape. It has two parameters: expectancy and a real number that makes it possible to calculate the probabilities of a group of random continuous variables.

Objective: this is a shortcut for talking about objective function. This function is the principal criterion for determining the solution of a mathematical optimization problem.

ODM (origin/destination model): macroscopic model of road traffic.

ODS (open document spreadsheet): file extension and open document format dedicated to spreadsheets based on the ODF (Open Document Format for Office Applications) standard. This format is used, in particular, by "Calc", the spreadsheet from the OpenOffice Suite.

OOP (object-oriented programming): computerized programming mode created in the 1960s by O.J. Dahl and K. Nygaard, then continued by A. Kay. It is built around objects that are software modules interacting with each other and representing an idea, a concept or an existing entity in the physical world. Numerous languages such as Java, C++, PHP and Ada using this programming mode. The concept of an object is also present within SE (Software Environments – Microsoft Visual Studio, Netbeans, etc.) in certain modeling tools (UML, DBDesigner, etc.) and in distributed buses (Corba, Pyro, RMI, etc.).

PEP (program evaluation procedure): planning method used by the United States Air Force.

PERT (program evaluation and review technology or **program evaluation research task):** project planning method developed by Booz, Allen and Hamilton, near the end of the 1950s, at the request of the Special Projects Office of the United States Navy. It makes it possible to create a connected, directed and weighted graph, where each of the tasks is represented by an arc that joins two vertices. It optimizes the project while taking into account the dependency constraints in order to determine the critical path and tasks.

Pivot: also known as the "Gauss Pivot". The pivot is attached to the method of the same name (also called the "Gauss–Jordan elimination"). This is an algorithm that determines the solutions of a system of linear equations.

Primal: refers to the initial problem in linear programming.

RBDMS (relational database management system): database management system that manages the relationships forming the constraints that guarantee the referential integrity of the data.

Relation: equivalent to an association between two entities in a relational model. Link established between two tables within an RBDMS.

Relational algebra: this is the logical and mathematical support on which the relational model is based. It offers a group of elementary operations for creating new relations.

Retroactive planning: also known as "feedback scheduling". This is an inverted planning method that conceives of a project starting with the end date and then go back toward the beginning. This method may be used when the end date of the project is determined in advance and mandatory.

SEP (Separation and Evaluation Procedure): see *Branch and bound*.

VBA (Visual Basic Application): implementation of the Microsoft Visual Basic language within Microsoft Office applications. Some applications such as WordPerfect, SolidWorks or AutoCAD also integrate a part of this language.

VBE (Visual Basic Editor): editing and development tool in the VBA language. This can also be an acronym for Visual Basic for Excel.

WBS (work breakdown structure): system of hierarchical breakdown of the tasks of a project.

Bibliography

[AHM 96] AHMED K.I., BEN-AKIVA M.E., KOUTSOPOULOS H.N. *et al.*, "Models of freeway lane changing and gap acceptance behavior", *Proceedings of the 13th International Symposium on the Theory of Traffic Flow and Transportation*, Lyon, France, 1996.

[BAN 95] BANDO M., "Dynamical model of traffic congestion and numerical simulation", *Physical Review E*, vol. 51, pp. 1035, 1995.

[BAR 10] BARCELO J., *Fundamentals of Traffic Simulation*, Springer, New York, 2010.

[BEN 14] BEN-AKIVA M., BIERLAIRE M., BURTON D. *et al.*, "Network state estimation and prediction for real-time transportation management applications", *Networks and Spatial Economics*, vol. 1, nos. 3–4, pp. 267–292, 2014.

[BEN 01] BEN-AKIVA M., KOUTSOPOULOS H.N., WALKER J., "DynaMIT-P. Dynamic assignement model system for transportation planning", *Proceedings of the 2001 World Conference in Transportation Research*, Seoul, Korea, 2001.

[BEN 98] BEN-AKVIA M., BIERLAIRE M., KOUTSOPOULOS H. *et al.*, DynaMIT: a simulation-based system for traffic prediction, DACCORD Short Term Forecasting Workshop, 1998.

[BIA 13] BIAFORE B., *Microsoft Project 2013: The Missing Manual*, 1st ed., O'Reilly Media, Sebastopol, CA, 2013.

[BIH 92] BIHAM O., MIDDLETON A.A., LEVINE D., "Self-organization and a dynamical transition in traffic-flow models", *Physical Review A*, vol. 46, no. 10, pp. R6124–R6127, 1992.

[BOL 02] BOLLOBAS B., *Modern Graph Theory*, Springer, New York, 2002.

[BRO 97] BRONSON R., *Schaum's Outline of Operations Research*, 2nd ed., McGraw-Hill, New York, 1997.

[CAS 95] CASSIDY M., MADANAT S.M., WANG M., "Unsignalized intersection capacity and level of service: revisiting critical gap", Transportation Research Record, 1995.

[CHA 58] CHANDLER R.E., HERMAN R., MONTROLL E.W., "Traffic dynamics: studies in car-following", *Operations Research*, vol. 6, pp. 165–184, 1958.

[CHO 06] CHOUDHURY C.F., BEN-AKIVA M.E., TOLEDO T., Cooperative lane changing and forced merging model, FHWA Draft Report, 2006.

[CLO 08] CLOUSE M., *Algèbre relationnelle: Guide pratique de conception d'une base de données relationnelle normalisée*, Editions ENI, Saint Herblain, 2008.

[COS 13] COSTESEQUE G., *Epistemologies and Practices of Modeling and Simulation*, Editions Matériologiques, France, 2013.

[CRE 86] CREMER M., LUDWIG J., "A fast simulation model for traffic flow on the basis of Boolean operations", *Mathematics and Computers in Simulation*, vol. 28, pp. 297–303, 1986.

[CRO 68] CROLAIS M., *Gestion intégrée de la production et ordonnancement*, Dunod, Paris, 1968.

[DAG 94] DAGANZO C., "The cell transmission model, part II: network traffic", *Transportation Research Part B: Methodological 28*, pp. 269–287, 1994.

[DAG 81] DAGANZO C., *Estimation of gap acceptance parameters within ans across the population from roadside observation*, Pergamon Press, 1981.

[DAG 95] DAGANZO, C., "Requiem for second-order fluid approximations of traffic flow", *Transportation Research Part B: Methodological 29*, pp. 79–93, 1995.

[DOR 91] DORNIER P.P., *Plein flux sur l'entreprise : La nouvelle logistique, de la gestion des stocks à la gestion des flux*, Nathan Entreprise, Paris, 1991.

[DRO 08] DROGOUL A., TREUIL J.P., ZUCKER J.D., *Modélisation et simulation à base d'agents : Exemples commentés, outils informatiques et questions théoriques*, Dunod, Paris, 2008.

[DUR 14] DURET A., CHIABAUT N., *Modélisation dynamique des congestions routières – Modèle de trafic LWR*, Techniques de l'ingénieur, 2014.

[FAU 96] FAURE R., *Précis de recherche opérationnelle*, 3rd ed., Dunod, Paris, 1996.

[GAR 05] GARDARIN G., *Base de données*, 6th ed., Eyrolles, Paris, 2005.

[GAW 98] GAWRON C., Simulation-based traffic assignment: computing user equilibria in large street networks, PhD Thesis, University of Cologne, Cologne, 1998.

[GAZ 61] GAZIS D.C., HERMAN R., ROTHERY R.W., "Nonlinear follow-the-leader models of traffic flow", *Operations Research*, vol. 9, no. 4, pp. 545–567, 1961.

[GIA 96] GIARD V., *Gestion de la production*, 2nd ed., Economica, Paris, 1996.

[GIP 81] GIPPS P.G., "A behavioural car-following model for computer simulation", *Transportation Research Part B*, vol. 15, pp. 105–111, 1981.

[GON 09] GONDRAN M., MINOUX M., *Graphes et algorithmes*, 4th ed., Lavoisier, Paris, 2009.

[GOR 15] GORDON R., *Intelligent Transportation Systems: Functional Design for Effective Traffic Management,* 2nd ed., Birkhauser Verlag AG, Switzerland, pp. 282, 2015.

[GRO 05] GROSS J.L., YELLEN J., *Graph Theory and Its Applications*, Chapman & Hall, Boca Raton, 2005.

[GOT 04] GOTHA G., *Modèles et algorithmes d'ordonnancement*, Ellipses, Paris, 2004.

[HID 02] HIDAS P., "Modeling lane changing and merging in microscopic traffic simulation", *Transportation Research Part C*, vol. 10, no. 5, pp. 351– 371, 2002.

[JIN 03] JIN W.-L., ZHANG H. M., "On the distribution schemes for computing fluxes through a highway merge", *Transportation Research Part B*, vol. 37, no. 6, pp. 521–540, 2003.

[KAL 16] KALA R., *On-Road Intelligent Vehicles: Motion Planning for Intelligent Transportation,* Butterworth-Heinemann, 2016.

[KIT 93] KITA H., "Effects of merging lane length on the merging behavior at expressway on-ramps", *Proceedings of the 12th International Symposium on the Theory of Traffic Flow and Transportation*, 1993.

[KOM 59] KOMETANI E., SASAKI T., "A safety index for traffic with linear spacing", *Operations Research*, vol. 7, no. 6, pp. 704– 720, 1959.

[KON 09] KONDYLI A., Breakdown probability model at freeway-ramp merges based on driver behavior, PhD Thesis, University of Florida, 2009.

[KOR 10] KORTE B., VYGEN J., *Optimisation combinatoire : Théorie et algorithmes*, Springer, Berlin, 2010.

[LAU 10] LAUGIÉ H., *VBA Excel 2010 – Créez des applications professionnelles: Exercices et corrigés*, Editions ENI, Saint Herblain, 2010.

[LE 15] LE GUEN F., *Macros et langage VBA – Apprendre à programmer sous Excel*, Editions ENI, Saint Herblain, 2015.

[LEC 02] LECLERC L., Modélisation dynamique du trafic et application à l'estimation du bruit routier, INSA, Lyon, 2002.

[LEB 05] LEBACQUE J.P., "Intersection modeling, application to macroscopic network traffic flow models and traffic management", *Traffic and Granular Flow* Part 3, pp. 261–278, Springer – France, 2005.

[LEO 89] LEONARD D.R., POWER P., TAYLOR N.B., CONTRAM: structure of the model, Transportation Research Laboratory, Crowthorn TRL Report RR 178, 1989.

[MAM 07] MAMMAR S., *Systèmes de transport intelligents: Modélisation, information et contrôle,* Hermes Science-Lavoisier, 2007.

[MAR 13] MARIAPPAN P., *Operations Research: An Introduction*, Pearson, London, 2013.

[MIC 89] MICHAELS R.M., FAZIO J., "Driver behavior model of merging", *Transportation Research Record*, vol. 1213, pp. 4–10, 1989.

[MIL 10] MILLER F.P., *Systèmes de transport intelligent : Technologies des l'information et de la communication, sécurité routière, réseau de télécommunication, urbanisation, infrastructure de transport,* Alphascript Publishing, France, 2010.

[NEW 02] NEWELL G.F., A Simplified Car-Following Theory: A Lower Order Model, Institute of Transportation Studies, University of California, Berkeley, 2002.

[NEW 61] NEWELL G.F., "Nonlinear effects in the dynamics of car-following", *Operations Research*, vol. 9, pp. 209–229, 1961.

[PIN 14] PINEDO M.L., *Scheduling: Theory, algorithms, and systems*, 4th ed., Springer, Heidelberg, 2014.

[PIP 53] PIPES L.A., "An operational analysis of traffic dynamics", *Journal of Applied Physics*, vol. 24, pp. 274–281, 1953.

[PRI 71] PRIGOGINE I., HERMAN R., *Kinetic Theory of Vehicular Traffic,* American Elsevier New-York, 1971.

[PRI 11] PRINS C., SEVAUX M., *Programmation linéaire avec Excel*, Eyrolles, Paris, 2011.

[REV 17a] RÉVEILLAC J.-M., *Modeling and Simulation of Logistics Flows 1*, ISTE, London and John Wiley & Sons, New York, 2017.

[REV 17b] RÉVEILLAC J.-M., *Modeling and Simulation of Logistics Flows 2*, ISTE, London and John Wiley & Sons, New York, 2017.

[ROB 11] ROBERT J., BERRY W., WHYBARK C. *et al.*, *Manufacturing Planning and Control for Supply Chain Management*, McGraw-Hill, Boston, 2011.

[SAW 13] SAWANT N., *Intelligent Transportation Systems: Longitudinal Vehicle Speed Controller for Autonomous Driving in Urban Stop-and-Go Traffic Situations,* Lambert Academic Publishing, USA, pp. 92, 2013.

[SNI 10] SNIEDOVICH M., *Dynamic Programming: Foundations and Principles*, 2nd ed., CRC Press, Boca Raton, 2010.

[STO 00] STOCK J., *Strategic Logistics Management*, 4th ed., McGraw-Hill, Boston, 2000.

[TOL 14] TOLBA C., *Les réseaux de Petri et leur utilisation pour la modélisation et la régulation du trafic routier*, Omniscriptum, Saarbrücken, 2014.

[TOL 03] TOLEDO T., Integrated driving behavior modeling, PhD Thesis, Massachusetts Institute of Technology, 2003.

[TRI 12] TRIEBER M., KESTING A., *Traffic Flow Dynamics: Data Models and Simulation*, Springer, Heidelberg, 2012.

[TRE 00] TREIBER M., HENNECKE A., HELBING D., "Congested traffic states in empirical observations and microscopic simulations", *Physical Review E*, vol. 62, no. 2, pp. 1805–1824, 2000.

[TRO 92] TROUTBECK R.J., Estimating the critical acceptance gap from traffic movements, Report, Physical Infrastructure Center, 1992.

[VAJ 13] VAJDA S., *Linear Programming: Algorithms and Applications*, Springer, Berlin, 2013.

[VAL 08] VALLIN P., *Problématique de la logistique*, Economica, Paris, 2008.

[VED 85] VÉDRINE J.P., *T.Q.G.*, Vuibert, Paris, 1985.

[VID 95] VIDAL C., *La recherche opérationnelle*, Presses Universitaires de France, Paris, 1995.

[WAL 13] WALKENBACH J., *Excel Power Programming with VBA*, Wiley, New York, 2013.

[WAN 05] WANG J., LIU R., MONTGOMERY F., "A simulation model for motorway merging behavior", *Proceedings of the 16th International Symposium on Transportation and Traffic Theory: Flow, Dynamics and Human Interaction*, University of Leeds, UK, 2005.

[WEN 11] WENG J., MENG Q., "Modeling speed-flow relationship and merging behavior in work zone merging area", *Transportation Research Part C19*, pp. 985–996, 2011.

[WIN 97] WINSTON W.L., *Operations Research Applications and Algorithms*, 3rd ed., Duxbury Press, 1997.

Internet links

Internet links are by their very nature subject to change. Over time, they might be moved to other addresses or may disappear. All the website provided here were valid at the time of this work going to print, but should any of them fail to work, a quick Google search should enable us to find them.

Graph theory and operational research

ELSEVIER: European journal of operational research *(in French)*, www.journals.elsevier.com/european-journal-of-operational-research/

INSA ROUEN: Graph theory and operational research *(in French)*, https://moodle.insa-rouen.fr/course/view.php?id=124

ROADEF: French society of operational research and decision aids *(in French)*, www.roadef.org/content/road/road.htm

University of Texas: Operations research models www.me.utexas.edu/~jensen/models/index.html

University of Nancy: Maximal flow in a graph *(in French)*, J.F. SCHEID www.iecn.u-nancy.fr/~scheid/Enseignement/flotmax.pdf

IEOR Berkeley: Network flows and graphs www.ieor.berkeley.edu/~ieor266/Lecture14.pdf

University of Metz: Modelling maximal flow problems *(in French)*, http://ensrotice.sciences.univ-metz.fr/module_avance_thg_voo6/co/modelflotmax.html

ENSTA: Introduction to discreet optimization *(in French)*, Adam OUOROU http://wwwdfr.ensta.fr/Cours/docs/MAE41/maxflow_hd.pdf

Tutorials on graphs, paths, trees, flows *(in French)* http://idmme06.inpg.fr/~rapinec/Graphe/Graphe/default.html

E-book: Applied Mathematical Programming, Bradley, Hax, and Magnanti (Addison-Wesley, 1977) http://web.mit.edu/15.053/www/

Article: On the history of the transportation and maximum flow problems, Alexander Schrijver http://homepages.cwi.nl/~lex/files/histtrpclean.pdf

Introduction to graph theory *(in French)*, Didier Müller www.apprendre-en-ligne.net/graphes/

CNRS: Graph theory and optimization in graphs *(in French)*, Christine Solnon http://liris.cnrs.fr/csolnon/polyGraphes.pdf

INPL: Elements of graph theory and linear programming *(in French)*, Didier Maquin http://cours.ensem.inpl-nancy.fr/cours-dm/graphes/Graphesnew.pdf

LABRI (LAboratoire Bordelais de Recherche en Informatique), Introduction to graph theory *(in French)*, Bruno Courcelle www.labri.fr/perso/courcell/Conferences/GraphesX.pdf

Graph theory *(in French)*, http://theoriedesgraphes.com/

Optimization and complexity: An extensive set of tutorials *(in French)*, Huet http://doc.ium.bz/S6/Optimisation %20et %20complexit %C3 %A9/cour/

GeoGebra: The calculating software for functions, geometry, algebra etc. Site for resources and downloads http://www.geogebra.org

GeoGebra: A tutorial written in English by Marc Renault http://webspace.ship.edu/msrenault/tutorial/

GeoGebra: New figures to see and make with GeoGebra *(in French)*, http://rdassonval.free.fr/geogebra/

Databases and relational algebra

University of Paris 8: Database, relational algebra *(in French)*, Rim Chaabane www.ai.univ-paris8.fr/~lysop/bd/seance5-ModeleRel-suite.pdf

ENST: The relational model and algebra *(in French)*, www.enst.dz/ Cours/ LeModele_relationneletAlgebre_relationnelle.pdf

Introduction to SQL language *(in French)*, Alexandre Mesle http://alexandre-mesle.com/enseignement/sql/index.html

What is an SGBDR? *(in French)*, Fabien Celaia http://fadace.developpez.com/sql/coddsgbdr/

Tables, Pivot Tables and VBA with Microsoft Excel

Excel practice, free VBA tutorial *(in French)* www.excel-pratique.com/fr/vba.php

Documentation for VBA developers, Microsoft *(in French)*, https://msdn.microsoft.com/fr-fr/office/ff688774.aspx

Excel exercise – Relations between tables *(in French)*, www.excel-exercice.com/excel2013/relation-entre-les-tables/

How to create a spreadsheet using Excel *(in French)*, WikiHow http://fr.wikihow.com/créer-un-tableau-avec-Excel

Paris-Dauphine University: VBA initiation for Excel *(in French)*, Philippe Bernard www.master272.com/finance/GP_L3/docs/VBA.pdf

Towson University: Excel 2013: PivotTables and Macros, Pamela J. Taylor https://www.towson.edu/adminfinance/ots/trainingdoc/customguide/Excel2013/Excel%202013_PivotTables%20and%20Macros_studentFINAL.pdf

College of Business Administration, Kansas State University: Conducting Data Analysis Using a Pivot Table, Brian Kovar http://info.cba.ksu.edu/bkovar/PivotTableTutorial.pdf

Spreadsheets in Excel 2007 *(in French)* http://silkyroad.developpez.com/excel/tableau/

AgroParisTech: Using lists of data with Excel 2007 *(in French)*, Michel Cartereau www.agroparistech.fr/mmip/mc/bazar/envoi.php?nom_fichier=cours-listes-excel 2007.pdf

Flow simulation

ExtendSim, flow simulation software, editor's website www.extendsim.com

A download of the demo version of ExtendSim https://www.extendsim.com/prods_demo.html

FlexSim, flow simulation software, editor's website https://www.flexsim.com

Arena, flow simulation software, editor's website https://www.arenasimulation. com

Witness, flow simulation software, editor's website www.lanner.com

SimWalk, simulation software for pedestrian flow, editor's website www.simwalk.com

PTV Group, PTV Vissim, traffic simulation software, editor's website www.ptvgroup.com

TSS, Aimsun, traffic simulation software, editor's website www.aimsun.com

Caliper, TransModeler, traffic simulation software, editor's website www.caliper.com/transmodeler/

AnyLogic, traffic simulation software, editor's website www.anylogic.com

Simio, flow simulation software, editor's website www.simio.com

Simul8, flow simulation software, editor's website www.simul8.com

SFPE, Society of Fire Protection http://www.sfpe-chapitre-francais.fr/ (French website) http://www.sfpe.org/ (American website)

Project management

Sciforma, project management software, editor's website www.sciforma.com

Visual Planning, project management software, editor's website http://www.visual-planning.com

Microsoft Project, project management software, editor's website https://products.office.com/fr-fr/project/project-and-portfolio-management-software

GanttProject, project management software, editor's website www.ganttproject.biz

The Wikipedia page devoted to project management software, including an enormous list of applications *(in French)*, http://fr.wikipedia.org/wiki/Logiciel_de_gestion_de_projets

Index

2D ANIMATION, 22
2D SYMBOL, 33
3D flows, 109

A, B, C

ACTIVITY, 81, 91, 93
ADD
  CONNECTION LINE
    ANIMATION, 22, 34
  SELECTED TO REPORT, 30
add-in, 44, 108
AGENTS, 125
analysis, 29
anchor point, 115
appearance, 40
ASTAR, 164
automated guided vehicle (AGV), 164
automated storage and retrieval system (ASRS), 162
BATCH, 16, 80, 83
BEHAVIORS, 140
Boids, 146
bottling, 50, 53, 66
BUTTON, 43
carbonation, 50, 81
card reader, 5, 10
chairlift, 3
change units, 48
CHANGE UNITS, 62
checks, 10
Clones, 42
constraints, 4, 48
continuous flows, 47
conversion, 49
CONVEY ITEM, 18, 34, 59
CONVEYORS, 50, 159, 163
CREATE, 66, 80, 83, 91, 93
CSV, 110
curves, 34

D, E ,F

discrete
  elements, 47
  flows, 47
elevators, 148
EMPIRICAL TABLE, 34
Engineers Society of Fire Protection (SFPE), 110, 146
evacuations, 109
exchange, 48, 68
EXECUTIVE, 6, 32, 53
EXIT, 21, 70, 83
experimenter, 231
ExtendSim, 1

FlexSim, 159
flows, 22
FLUID, 164

G, H, I

GENERATE REPORT, 29, 31
generators, 7
GLOBAL TIME UNIT, 6, 53
heating, 49
  conduit, 49
hierarchical block, 38
hierarchy, 38
INTERCHANGE, 66
Item.lix, 53, 66, 80, 99

L, M, N

Levant, 1
libraries, 162
LOOKUP TABLE, 75
maintenance, 51, 93
MAX & MIN, 13
MAXIMUM OUTFLOW RATE, 66
MERGE, 57, 65, 119
MINIMUM VALUE, 14
mix, 49
mixed flows, 47
model design, 7
NEW STRING ATTRIBUTE, 32

O, P, Q

OPEN STRUCTURE, 39
opening times, 7, 9, 48, 53, 94
Pathfinder, 109
PERSON_M, 22
plan, 114
plastic film, 50
PLOTTER, DISCRETE EVENT, 34, 70
POOL, 25, 93
principal parameters, 5
PROPERTY VALUE, 31, 33
PyroSim, 110
QUEUE, 11, 67, 81

R, S

RANDOM NUMBER, 31
Rate.lix, 55, 66
repetitive testing, 98
report, 29
RESOURCE POOL
  QUEUE, 25
  RELEASE, 27, 94, 97
resources, 24, 93
RHEOSTAT, 74
Running, 22
Scitor Process, 1
SELECT ITEM
  IN, 11, 20
  OUT, 11, 20
setting up the project, 5
SHOW BLOCK NUMBERS, 32
SHUTDOWN, 99
simulation of flows, 1
SIMULATION SETUP, 8, 54
ski lift, 3
soda, 48
specifications, 3
station, 2
STATISTICS, 91, 107
steering mode, 146
subjective view, 139
swipe card, 5

T, V, W

TANK, 55, 61
TEXT, 41
TRANSPORT, 86
UNBATCH, 18
Value.lix, 91
VALVE, 57, 64
warehousing, 159
weight-volume, 49

Other titles from

ISTE

in

Systems and Industrial Engineering – Robotics

2016

ANDRÉ Michel, SAMARAS Zissis
Energy and Environment
(Research for Innovative Transports Set - Volume 1)

AUBRY Jean-François, BRINZEI Nicolae, MAZOUNI Mohammed-Habib
Systems Dependability Assessment: Benefits of Petri Net Models (Systems Dependability Assessment Set - Volume 1)

BLANQUART Corinne, CLAUSEN Uwe, JACOB Bernard
Towards Innovative Freight and Logistics (Research for Innovative Transports Set - Volume 2)

COHEN Simon, YANNIS George
Traffic Management (Research for Innovative Transports Set - Volume 3)

MARÉ Jean-Charles
Aerospace Actuators 1: Needs, Reliability and Hydraulic Power Solutions

REZG Nidhal, HAJEJ Zied, BOSCHIAN-CAMPANER Valerio
Production and Maintenance Optimization Problems: Logistic Constraints and Leasing Warranty Services

TORRENTI Jean-Michel, LA TORRE Francesca
Materials and Infrastructures 1 (Research for Innovative Transports Set - Volume 5A)
Materials and Infrastructures 2 (Research for Innovative Transports Set - Volume 5B)

WEBER Philippe, SIMON Christophe
Benefits of Bayesian Network Models
(Systems Dependability Assessment Set – Volume 2)

YANNIS George, COHEN Simon
Traffic Safety (Research for Innovative Transports Set - Volume 4)

2015

AUBRY Jean-François, BRINZEI Nicolae
Systems Dependability Assessment: Modeling with Graphs and Finite State Automata

BOULANGER Jean-Louis
CENELEC 50128 and IEC 62279 Standards

BRIFFAUT Jean-Pierre
E-Enabled Operations Management

MISSIKOFF Michele, CANDUCCI Massimo, MAIDEN Neil
Enterprise Innovation

2014

CHETTO Maryline
Real-time Systems Scheduling
Volume 1 – Fundamentals
Volume 2 – Focuses

DAVIM J. Paulo
Machinability of Advanced Materials

ESTAMPE Dominique
Supply Chain Performance and Evaluation Models

FAVRE Bernard
Introduction to Sustainable Transports

GAUTHIER Michaël, ANDREFF Nicolas, DOMBRE Etienne
Intracorporeal Robotics: From Milliscale to Nanoscale

MICOUIN Patrice
Model Based Systems Engineering: Fundamentals and Methods

MILLOT Patrick
Designing Human–Machine Cooperation Systems

NI Zhenjiang, PACORET Céline, BENOSMAN Ryad, REGNIER Stéphane
Haptic Feedback Teleoperation of Optical Tweezers

OUSTALOUP Alain
Diversity and Non-integer Differentiation for System Dynamics

REZG Nidhal, DELLAGI Sofien, KHATAD Abdelhakim
Joint Optimization of Maintenance and Production Policies

STEFANOIU Dan, BORNE Pierre, POPESCU Dumitru, FILIP Florin Gh.,
EL KAMEL Abdelkader
Optimization in Engineering Sciences: Metaheuristics, Stochastic Methods andDecision Support

2013

ALAZARD Daniel
Reverse Engineering in Control Design

ARIOUI Hichem, NEHAOUA Lamri
Driving Simulation

CHADLI Mohammed, COPPIER Hervé
Command-control for Real-time Systems

DAAFOUZ Jamal, TARBOURIECH Sophie, SIGALOTTI Mario
Hybrid Systems with Constraints

FEYEL Philippe
Loop-shaping Robust Control

FLAUS Jean-Marie
Risk Analysis: Socio-technical and Industrial Systems

FRIBOURG Laurent, SOULAT Romain
Control of Switching Systems by Invariance Analysis: Application to Power Electronics

GROSSARD Mathieu, REGNIER Stéphane, CHAILLET Nicolas
Flexible Robotics: Applications to Multiscale Manipulations

GRUNN Emmanuel, PHAM Anh Tuan
Modeling of Complex Systems: Application to Aeronautical Dynamics

HABIB Maki K., DAVIM J. Paulo
Interdisciplinary Mechatronics: Engineering Science and Research Development

HAMMADI Slim, KSOURI Mekki
Multimodal Transport Systems

JARBOUI Bassem, SIARRY Patrick, TEGHEM Jacques
Metaheuristics for Production Scheduling

KIRILLOV Oleg N., PELINOVSKY Dmitry E.
Nonlinear Physical Systems

LE Vu Tuan Hieu, STOICA Cristina, ALAMO Teodoro,
CAMACHO Eduardo F., DUMUR Didier
Zonotopes: From Guaranteed State-estimation to Control

MACHADO Carolina, DAVIM J. Paulo
Management and Engineering Innovation

MORANA Joëlle
Sustainable Supply Chain Management

SANDOU Guillaume
Metaheuristic Optimization for the Design of Automatic Control Laws

STOICAN Florin, OLARU Sorin
Set-theoretic Fault Detection in Multisensor Systems

2012

AÏT-KADI Daoud, CHOUINARD Marc, MARCOTTE Suzanne, RIOPEL Diane
Sustainable Reverse Logistics Network: Engineering and Management

BORNE Pierre, POPESCU Dumitru, FILIP Florin G., STEFANOIU Dan
Optimization in Engineering Sciences: Exact Methods

CHADLI Mohammed, BORNE Pierre
Multiple Models Approach in Automation: Takagi-Sugeno Fuzzy Systems

DAVIM J.Paulo
Lasers in Manufacturing

DECLERCK Philippe
Discrete Event Systems in Dioid Algebra and Conventional Algebra

DOUMIATI Moustapha, CHARARA Ali, VICTORINO Alessandro, LECHNER Daniel
Vehicle Dynamics Estimation using Kalman Filtering: Experimental Validation

GUERRERO José A, LOZANO Rogelio
Flight Formation Control

HAMMADI Slim, KSOURI Mekki
Advanced Mobility and Transport Engineering

MAILLARD Pierre
Competitive Quality Strategies

MATTA Nada, VANDENBOOMGAERDE Yves, ARLAT Jean
Supervision and Safety of Complex Systems

POLER Raul *et al.*
Intelligent Non-hierarchical Manufacturing Networks

TROCCAZ Jocelyne
Medical Robotics

YALAOUI Alice, CHEHADE Hicham, YALAOUI Farouk, AMODEO Lionel
Optimization of Logistics

ZELM Martin *et al.*
Enterprise Interoperability –I-EASA12 Proceedings

2011

CANTOT Pascal, LUZEAUX Dominique
Simulation and Modeling of Systems of Systems

DAVIM J. Paulo
Mechatronics

DAVIM J. Paulo
Wood Machining

GROUS Ammar
Applied Metrology for Manufacturing Engineering

KOLSKI Christophe
Human–Computer Interactions in Transport

LUZEAUX Dominique, RUAULT Jean-René, WIPPLER Jean-Luc
Complex Systems and Systems of Systems Engineering

ZELM Martin, *et al.*
Enterprise Interoperability: IWEI2011 Proceedings

2010

BOTTA-GENOULAZ Valérie, CAMPAGNE Jean-Pierre, LLERENA Daniel, PELLEGRIN Claude
Supply Chain Performance / Collaboration, Alignement and Coordination

BOURLÈS Henri, GODFREY K.C. Kwan
Linear Systems

BOURRIÈRES Jean-Paul
Proceedings of CEISIE'09

CHAILLET Nicolas, REGNIER Stéphane
Microrobotics for Micromanipulation

DAVIM J. Paulo
Sustainable Manufacturing

GIORDANO Max, MATHIEU Luc, VILLENEUVE François
Product Life-Cycle Management / Geometric Variations

LOZANO Rogelio
Unmanned Aerial Vehicles / Embedded Control

LUZEAUX Dominique, RUAULT Jean-René
Systems of Systems

VILLENEUVE François, MATHIEU Luc
Geometric Tolerancing of Products

2009

DIAZ Michel
Petri Nets / Fundamental Models, Verification and Applications

OZEL Tugrul, DAVIM J. Paulo
Intelligent Machining

PITRAT Jacques
Artificial Beings

2008

ARTIGUES Christian, DEMASSEY Sophie, NERON Emmanuel
Resources–Constrained Project Scheduling

BILLAUT Jean-Charles, MOUKRIM Aziz, SANLAVILLE Eric
Flexibility and Robustness in Scheduling

DOCHAIN Denis
Bioprocess Control

LOPEZ Pierre, ROUBELLAT François
Production Scheduling

THIERRY Caroline, THOMAS André, BEL Gérard
Supply Chain Simulation and Management

2007

DE LARMINAT Philippe
Analysis and Control of Linear Systems

DOMBRE Etienne, KHALIL Wisama
Robot Manipulators

LAMNABHI Françoise *et al.*
Taming Heterogeneity and Complexity of Embedded Control

LIMNIOS Nikolaos
Fault Trees

2006

FRENCH COLLEGE OF METROLOGY
Metrology in Industry

NAJIM Kaddour
Control of Continuous Linear Systems

Printed and bound by CPI Group (UK) Ltd, Croydon, CR0 4YY
16/09/2026
14963072-0004